TOWARDS TRANSPORT NET ZERO

TRANSPORT AND SUSTAINABILITY

Series Editors: Stephen Ison, Jon Shaw and Maria Attard

Recent Volumes:

Volume 1:	Cycling and Sustainability
Volume 2:	Transport and Climate Change
Volume 3:	Sustainable Transport for Chinese Cities
Volume 4:	Sustainable Aviation Futures
Volume 5:	Parking: Issues and Policies
Volume 6:	Sustainable Logistics
Volume 7:	Sustainable Urban Transport
Volume 8:	Paratransit: Shaping the Flexible Transport Future
Volume 9:	Walking: Connecting Sustainable Transport with Health
Volume 10:	Transport, Travel and Later Life
Volume 11:	Safe Mobility: Challenges, Methodology and Solutions
Volume 12:	Urban Mobility and Social Equity in Latin America: Evidence, Concepts, Methods
Volume 13:	Sustainable Transport and Tourism Destinations
Volume 14:	Sustainable Railway Engineering and Operations
Volume 15:	Electrifying Mobility: Realising a Sustainable Future for the Car
Volume 16:	Women, Work and Transport
Volume 17:	Transport and Pandemic Experiences
Volume 18:	Public Participation in Transport in Times of Change
Volume 19:	Sustainable Automated and Connected Transport

Emerald Publishing Limited
Emerald Publishing, Floor 5, Northspring, 21-23 Wellington Street, Leeds LS1 4DL.

First edition 2025

Printed and bound by CPI Group (UK) Ltd, Croydon, CR0 4YY

British Library Cataloguing in Publication Data
A catalogue record for this book is available from the British Library

ISBN: 978-1-83797-694-2 (Print)
ISBN: 978-1-83797-693-5 (Online)
ISBN: 978-1-83797-695-9 (Epub)

ISSN: 2044-9941 (Series)

This book is dedicated to Professor Donald Shoup, a world renowned scholar in the area of parking management.

CONTENTS

Tables

Boxes

ABOUT THE EDITORS

Jon Shaw is Professor of Transport Geography and Head of the School of Geography, Earth and Environmental Sciences at the University of Plymouth. He is the author of *Competition, Regulation and the Privatisation of British Rail*, and co-author of *Diverging Mobilities* and *The Transport Debate*. This is his seventh co-edited book. His research has focussed mainly on transport and mobility governance, policy, and strategy. He has served on the editorial boards of several journals, including as the Associate Editor of the *Journal of Transport Geography*, and has been a Specialist Adviser to the UK House of Commons' Transport Select Committee. He is a member of the Great Western Railway Stakeholder Advisory Board and Chair of the Devon and Cornwall Rail Partnership.

Stephen Ison is Professor of Transport Policy within Leicester Castle Business School at De Montfort University and Honorary Professor of Transport Policy at the Institute of Transport and Logistics Studies, University of Sydney. He has published widely in the area of economics and transport policy, including transport and sustainability, regulation and deregulation, and the use of the market-mechanism for addressing transport externalities. He has edited, authored, or co-authored 30 books, and published over 150 peer-reviewed journal papers. He is a member of the Scientific Committee of the World Conference on Transport Research, Founding Editor of the *Journal Research in Transportation Business and Management*, Co-editor-in-Chief of the *Journal of Transport and Sustainability*, and Book Series Co-editor of *Transport and Sustainability*.

Maria Attard is a Full Professor of Transport Geography at the University of Malta. She is also the Director of the Institute for Climate Change and Sustainable Development. She has published in the fields of urban transport and sustainable mobility, public transport, and active travel. She was a Consultant for Malta's government between 2002 and 2010 and implemented several projects in and around Valletta, the islands' capital. She holds editorial positions in a number of journals and a book series and supports numerous editorial boards. She is the current President of NECTAR, a European network of transport researchers, and sits on several international research committees in Europe and beyond.

ABOUT THE CONTRIBUTORS

Eda Beyazit is a Research Fellow in Active Travel and Micro-mobility at the Centre for Transport & Society (CTS), University of the West of England, Bristol. Before joining CTS in 2023, she worked as an Associate Professor of Urban Planning at the Faculty of Architecture, Istanbul Technical University, where she founded the IstanbulON Urban Mobility Lab. Her interests include transport-related inequalities, gendered mobilities, informality, and the urban periphery. At CTS, she works on projects covering a range of topics from rural transport and regional disparities, governance of transport systems, and health-related impacts of transport to stakeholder participation and engagement in transport.

Simon Blainey is Professor of Sustainable Transport at the University of Southampton. His background is in transport geography, and he has almost two decades' experience in research relating to railway systems, with a focus on rail demand modelling and appraisal and on cost and carbon modelling for railway infrastructure. More broadly, his research focusses on developing tools that help support decision-makers in planning sustainable transport systems that are appropriate to the geographic context in which they are to be delivered and which take full account of interdependencies with other elements of the engineered, social, and natural environments. He is particularly interested in using geospatial data and models to analyse transport systems, and during his career, this has involved work relating to nearly all major transport modes.

Clemens Brauer studied Industrial Engineering with a focus on environmental management and technology in Berlin. In 2008, he started employment in the automotive industry with a focus on simulating the effects of vehicle parameters and driving behaviour on fuel consumption and CO_2 emissions. As a volunteer, he led the 'Safe Energy and Water Supply for Two Educational Institutions in Zimbabwe' project for the Stuttgart Regional Group of *Ingenieure ohne Grenzen*. From 2017 to 2019, he worked as a consultant for digitalisation and mobility in Stuttgart, and since then has been a Research Associate at Fraunhofer ISI. His work focusses on data analysis and modelling in the field of sustainable mobility and transport systems.

Lucy Budd is a Human Geographer and Professor of Transport Management at Leicester Castle Business School, De Montfort University and Honorary Professor at the Institute of Transport and Logistics Studies, University of Sydney. Her current research includes projects on sustainable urban transport, micro-mobility, transport decarbonisation, and transport accessibility for people living with disabilities. She is Co-Editor-in-Chief of *Research in Transportation Business and Management* and Founding Co-Editor of the *Journal of Transport*

Graham Parkhurst is Professor of Sustainable Mobility and Director of the Centre for Transport & Society at the University of the West of England, Bristol. He has taught and researched topics related to transport policy since 1991. His current research examines the sustainable mobility implications of greater automation and electrification, and the emergence of new business models, taking a critical lens to the discourses and practices of 'future mobility' on land and in the air. In 2022, Emerald published his co-edited volume *Electrifying Mobility: Realising a Sustainable Future for the Car*. He is a Fellow of the Royal Geographical Society.

John Parkin is Emeritus Professor of Transport Engineering at the University of the West of England, Bristol. He worked for consulting engineers before joining academia and has been involved in all stages of the promotion of transport schemes including policy formulation, modelling and forecasting, planning, scheme appraisal and operational analysis, design and construction, and evaluation. With emphasis on human-scale travel, his research is at the nexus of user behaviour and infrastructure design. He authored *Designing for Cycle Traffic* and edited *Cycling and Sustainability* and *Cycling Futures: From Research into Practice*. He was a member of the Welsh Government Roads Review Panel and Commissioner on the North Wales Transport Commission.

John Preston is Professor of Rail Transport at the School of Engineering, University of Southampton, having previously held posts at the Universities of Leeds and Oxford. He has over 40 years of experience in transport research and education, having held almost 160 research grants and contracts worth over £23 million, published over 400 articles, book chapters, conference and working papers, and successfully supervised over 40 doctoral students. He is a Fellow of the Chartered Institution of Highways and Transportation, and in 2024, he was a member of the first inductee cohort to the Universities' Transport Study Group Pantheon Hall of Fame.

Tom Rye is a Professor of Transport Policy at Molde University College – Specialized University in Logistics, and a Principal Researcher in the Transformative Transport Planning team at the Urban Planning Institute of Slovenia. His research and consultancy projects over many years for local and national governments, research funding councils, and international institutions have often led him to consider the importance of public participation in project planning and implementation. He is currently Principal Investigator of the Norwegian research council project TRANSPOL, on politicians, public consultation, and contentious transport measures and, with the other authors of this chapter, co-edited *Public Participation in Transport in Times of Change*.

Tim Ryley joined Griffith University in 2015 as the inaugural Professor of Aviation. He has over 25 years of transport research experience across UK and Australian universities. His research has typically been industry linked, multi-disciplinary, and problem based. Collaborations have been with a wide range of industry and government partners, including major airports and airlines. His research has increasingly focussed on the environmental aspects of aviation and is an internationally leading expert in decarbonising the air transport industry.

Donald Shoup sadly passed away during the production of this volume. He was Distinguished Research Professor in UCLA's Department of Urban Planning. His research focussed on how parking policies affect cities, the economy, and the environment. He was a Fellow of the American Institute of Certified Planners and an Honorary Professor at the Beijing Transportation Research Center. The American Planning Association gave him its National Excellence Award for a Planning Pioneer, and the American Collegiate Schools of Planning gave him its Distinguished Educator Award.

Claus Hedegaard Sørensen is a Senior Researcher at VTI, The Swedish Road and Transport Research Institute, working within the 'Mobility, Actors and Planning' research unit. He is also connected to K2, The Swedish Knowledge Centre for Public Transport. A main focus of his research is the transformation of the transport sector to deliver a more sustainable society, while another main focus is transport governance, that is institutions, processes, policy instruments, and policy legitimacy. Often, his work combines these fields. Over the past few years, he has had a particular interest in how to gain legitimacy for transformative policies in the transport sector. He is Co-Editor of *Public Participation in Transport in Times of Change* (Emerald, 2023) and *Shaping Smart Mobility Futures* (Emerald, 2020).

Lóránt Tavasszy is Professor in Freight Transport and Logistics at Delft University of Technology. He studied Civil Engineering in Delft and has been with the Dutch applied research institute TNO in different roles between 1996 and 2016. His research and teaching concern freight transport system models and innovations. Besides the textbook *Modelling Freight Transport*, he has published widely on different innovative freight transport models at local, national, and global levels.

Maria A. Wilches-Mogollon is a Research Associate at the Bartlett Development Planning Unit at University College London, with over eight years of research experience. Her work explores the intersection between urban transport interventions and the impact on urban health, especially in low- and middle-income countries' cities and how to assess transport transformation for improving health indicators.

Luca Zamparini is Associate Professor of Economics and currently chairs the Tourism Economics course, at the Faculty of Economics of the University of Salento. He is also Cluster Coordinator (Cluster 5: Leisure, Recreation, and Tourism) of NECTAR and Editor of the Edward Elgar series *Comparative Perspectives on Transportation Security*. His main fields of expertise are transport and tourism economics with an emphasis on the qualitative attributes of transport services, transport security, from both theoretical and empirical viewpoints, and the relationships between transport and tourism development.

Yuting Zhang is a Research Fellow in Sustainable Mobility at the Institute of Transport and Logistics Studies at the University of Sydney. She has an urban planning background with a wide range of educational and professional experience in China, the United States, and Australia. Her research focusses on promoting active travel, such as walking and cycling, as well as evaluating the role of micro-mobility vehicles like e-scooters in supporting sustainability within the broader transportation network in urban areas. Proficient in policy development, spatial analysis using ArcGIS, and industrial collaborations, she is dedicated to contributing insights into the planning and implementation of sustainable transport solutions.

ACKNOWLEDGEMENTS

In founding and editing the *Transport and Sustainability* series, we owe a sincere debt of gratitude to a number of people. At Emerald, the entire team is due a huge 'thankyou' and we'd especially like to mention, from over the years, Cristina Irving Turner, Jen McCall, Nick Wolterman, Amber Stone-Galilee, Iram Satti, David Mulvaney, Daniel Ridge, and Lauren Kammerdiener. We are grateful to the Editorial Advisory Board members for their guidance, and it goes without saying that the Volume Editors have made an enormous contribution to the success of the series. They developed the initial ideas for their edited books and brought together the contributors necessary to sustain it over the years and, now, 20 volumes. It is maybe unwise to single out individual editors, although we would like to make special mention of John Parkin, who was the first to show faith in the series, and Corinne Mulley and Lucy Budd, who, respectively, co-edited four and two of the following volumes. And to each and every one of the more than 500 contributors: heartfelt thanks for your analysis and insight that brings each of the volumes in the series to life.

Specifically in relation to this volume, we would like to thank the contributors for writing their excellent chapters in the context of many other demands on their time (not to mention tedious emails from us). Jamie Quinn, of the Geomapping Unit at the University of Plymouth, expertly drew all of the artwork, also in the context of many other demands on his time, and at Emerald, Brindha Thirunavukkarasu and Lydia Cutmore helped us navigate the 'final mile'. Jon would additionally like to thank Paul Simpson, Ralph Fyfe and Chris Fogwill for their sage advice on keeping up a REF submission while doing fun things like edited books. Alex Belton, Richard Preziosi, Adrian Dawson and Dafydd Moore provided much needed strategic discussion. *Pisces non sunt animalia.* Without Donella Bone's fantastic assistance, the diary contortions necessary to fit all of this in simply wouldn't have been possible.

Jon Shaw, Stephen Ison and Maria Attard, May 2025.

the evening before a PhD viva, and Maria officially joined the team a couple of years later. All three of us wanted to initiate a series to fill what we thought was a gap in an area that linked the essence of our own research work and an expanding area of research in general. We saw the potential for a series that brought together the widest range of perspectives on issues and policies at the interface between transport and sustainability. Our aims were for the series to be international, inclusive, and based on cutting-edge research that was rigorously peer reviewed. It would be a collection of edited rather than sole-authored books, precisely to increase the range of contributors and thus perspectives. We were delighted that Emerald shared our vision for the series, which has now been in existence for almost a decade and a half.

Fifteen years after we began sketching out our proposal (quite literally in the sense of completing the necessary paperwork) we are proud of how the series has progressed; it now represents a substantial body of work. The 19 volumes preceding this one address transport and sustainability issues by mode (walking, cycling, railways, aviation), policy area (safety, paratransit, parking, participation, response to pandemic disruption), demographic considerations (gender, older people), and economic sector (logistics, tourism), as well as geographically (urban areas, China, Latin America) and technologically (electric vehicles, connected and autonomous vehicles (AVs)). It was also important to dedicate an entire volume early in the series to the relationship between transport and climate change.

The contributors are academics and practitioners from a wide range of fields, spanning the career stages of their respective professions. All told, 38 different editors have secured the contributions of more than 500 authors, 40% of whom are women, based in 37 countries on all five inhabited continents. There have been nearly 112,000 downloads of these contributions, again from all five continents. Around 30% have come from north and west Europe, with China, Australasia, South and East Europe, East Asia and North America accounting for the next largest proportions. At the time of writing, the most downloaded volume is *Sustainable Logistics* (over 15,000), and the most downloaded chapter is *The Health Benefits of Walking* (nearly 2,400 – around a quarter, or almost a half, depending on which study you believe, of your recommended daily number of steps!).

Not only is this volume the 20th in the *Transport and Sustainability* series, but it is also the first as Series Editors we have edited together. We thought it fitting to choose a theme that was globally significant and, to mark the milestone the 20th volume represents, could be explored through analyses relating to topics covered by the other 19 volumes in the series. Of all the sustainability grand challenges faced by humanity today, net zero stood out. Transport is particularly acutely associated with net zero and climate change, because of the sheer significance and apparently unrelenting nature of its contribution to overall carbon emissions as a sector. Globally, transport generates more than eight gigatonnes of CO_2 emissions each year – around a quarter of all greenhouse gas emissions – and almost all of the sector relies heavily on fossil fuels; fully 91% of its final energy use came from oil products in 2022 (International Energy Agency (IEA), 2023). Before the COVID-19 pandemic, transport was the largest source of energy-related greenhouse gas emissions in 45 countries, and everywhere else it was the second largest (United Nations (UN), 2021). With growth in the sector predicted to be high – one recent study suggests a doubling of pre-pandemic activity levels by 2050 (Tjandra et al., 2024) – swift and radical action, technologically, culturally, and behaviourally, is needed to decarbonise the ways we move people and goods around.

In this context, we invited editors and contributors from the previous volumes to write chapters that related to net zero from the point of view of the themes explored in their own books. What results is a significant and novel contribution to debates about the way in which the transport sector can – and must – navigate its way towards net zero. Since it focuses exclusively upon topics covered elsewhere in the series, our approach is not intended to be comprehensive. That said, in curating *Transport and Sustainability* as a wide-ranging series, we always had in mind the possibility of producing a book of this nature, providing the longevity of the series allowed. Most of the key themes are as such addressed in their own chapter, and others are discussed in the context of related topics at various places in the text (the need for a just decarbonisation transition and the relationship between decarbonisation and broader notions of sustainability are good examples). We thus present this volume as a celebration of the *Transport and Sustainability* series and its contributors, as well as a credible and authoritative source of information and insight on transport and decarbonisation. We summarise this information and insight in the next section of this chapter.

OVERVIEW OF THIS BOOK

Chapter 2 sets the scene by explaining why decarbonising the transport sector is so important in tackling climate change. Tim Ryley establishes key trends in global transport demand and related emissions, providing more details on the extent of the sector's contribution to total global carbon emissions. He notes that 'most governments and many non-state actors, such as companies and educational institutions, have committed to being net zero by 2050', and importantly in the context of the inexorable rise in transport's carbon emissions over recent decades, that 'addressing the CO_2 emissions of transport activities is a major component of this commitment'. While road transport still accounts for the largest share of the sector's CO_2 emissions at 74%, it is air transport that has witnessed the most significant growth (115% above 1990 levels). In this context, Tim explains that any progress towards a net zero emission pathway is complex and fraught with difficulty, and he identifies five major issues that are especially germane to any meaningful decarbonisation effort. The first three relate to mitigation: the need to reduce CO_2 emissions from the multitude of transport activities across different spatial scales, making swift use of technology where appropriate, and recognising the need for behaviour change. The fourth relates to adaptation. With 2024 confirmed as the hottest year on record and Earth's temperature 'clearly exceeding' 1.5 degrees above pre-industrial levels (Copernicus Programme, 2024, unpaginated), the transport industry needs to be 'planned and ready' for disturbances – more powerful storm events, flooding, wind damage, and excess heat – associated with climate change that has already happened and looks set to continue at pace. Finally, he calls for governments and the transport sector alike to be mindful of the need to ensure moves towards transport net zero are socially just, a theme that recurs throughout this book.

Chapters 3 and 4 focus on walking and what John Parkin refers to as the 'human-scale' modes. In their analysis of walking for net zero, Yuting Zhang and Corinne Mulley emphasise the significance of walking as a means not only of lessening dependence on fossil fuels but also of realising a host of social and personal health benefits. They argue that building robust walking infrastructure – accounting for the built environment, location, user characteristics, and enjoyment in its design – is a necessary but not sufficient component of the policymakers' toolkit. Such interventions not only need to be made on the basis of a detailed understanding of relevant data, but they also need to take place alongside sustained efforts to cultivate a culture of walking. Future societies in which walking is ingrained as a more popular way of completing many more different types of journey will be fitter, healthier, and more socially just, as well as further down the path towards transport net zero.

John in Chapter 4 is concerned essentially with small, wheeled vehicles – different kinds of cycles, e-bikes, scooters, and e-scooters. He sets out in the first part of the chapter lifecycle CO_2 equivalent (CO_2eq)[1] emissions from these human-scale mobility modes as compared with other modes (generally far less than vehicles with internal combustion engines (ICEs), although the comparisons with battery electric vehicles, especially in relation to shared human-scale mobility schemes, might surprise you), before moving on to discuss the nature of strategies and policies needed to promote their use. John's conclusion that promoting human-scale mobility is a prerequisite for any successful net zero transport agenda is clear. So too is his view that well-designed infrastructure is delivered in the context of a wide-ranging and logically consistent approach, from legislative underpinnings, through strategy and policy, to delivery on the ground. He refers to the Welsh Government's Future Generations (Wales) Act of 2015 that 'sets out requirements for improving the social, economic, environmental and cultural well-being of Wales and its citizens' as an example of how this is being attempted in practice.

Next, Graham Parkhurst provides the first of three chapters that deal largely (but not exclusively) with personal vehicles. In his detailed discussion of the transition to electric cars, Graham highlights key areas of progress (which is heartening) alongside three significant challenges that are proving resilient and not entirely straightforward to overcome (which is less so). In short, while take-up is increasing, it is geographically patchy and, in any event, technical difficulties still need to be overcome to produce *genuinely* net zero electric cars. Assuming this becomes achievable, we encounter cultural barriers because certain ease-of-use characteristics associated with an automobility regime that has grown up around ICE vehicles are awkward to replicate with electric equivalents. There are also problems in realising an 'attractive, affordable and profitable' environment for the mass take-up of electric vehicles for a host of reasons. Encouragingly, breakthroughs are being made on a number of these fronts, and in a world where there are still around 1 billion conventional ICE vehicles, a stringent policy approach '[b]ringing forward the point at which new ICEs no longer enter the global fleet is ... important'.

Eda Beyazit details in Chapter 6 the opportunities and challenges for net zero transport afforded by AV technology. Much hype has surrounded the introduction of AVs, not least because of their potential to 'free up' both drivers from the burden of having to control their vehicles and car-centric automobility systems from the burden of human drivers. There are indeed many potential benefits of AVs, including enhanced efficiency – computers are able to overcome driver unpredictability to the point where, other things being equal, it becomes possible to reduce congestion – as well as better safety and increased social inclusion. Positive net zero effects can also arise, especially when AVs are electric.[2] At the same time, realising significant net zero and congestion reduction benefits is a long-term aspiration because it depends on high market penetration rates, and there is the distinct risk that an automobility system based predominantly on AVs could materially increase vehicle kilometres travelled.

One reason for this is they can obviate the need to park once drivers (passengers!) have reached their destination. Why pay to leave one's car nearby when it can be sent home and summoned to return on demand? It is to the subject of parking that we turn in Chapter 7. Donald Shoup proposes parking reforms 'to improve cities and cool the planet'. He begins with the thought-provoking point that '[c]ars are parked for 95% of their lives, so the rubber meets the road mainly in parking spaces. Parked cars don't produce any carbon emissions, so parking reforms might seem a poor way to reduce carbon emissions'. You might expect Donald is setting up a straw person and you would be right: he goes on to advocate eight market-based initiatives, all of which have been adopted by city governments in California and beyond. The underlying idea is to transfer more of the cost of parking to parkers themselves and, in the process, 'make cities more expensive for cars and better for people'.

The next four chapters deal with shared transport modes – flexible transport services, railways, aviation, and maritime transport. In Chapter 8, Corinne Mulley and John D. Nelson focus on those passenger services that 'are flexible on one or more of the dimensions of the route: vehicle allocation, vehicle operator, type and time of payment and passenger category'. They illustrate their discussion with reference to eight issues that need to be addressed if flexible transport schemes (FTSs) are to be successful, ranging from the use of technology to the role of marketing the type of provision that is on offer. Their wide-ranging contribution discusses various examples around the world. While such schemes have the potential to act as a substitute for private car journeys in certain circumstances, we can't lose sight of the fact that a fundamental reason for their existence is to promote mobility and social inclusion among disadvantaged groups. Greening the vehicle fleet, especially in developing countries, becomes an especially important policy goal as a result.

Chapter 9 focuses on railways. In relation to other modes, train travel is often held up as something of a poster child of low-carbon transport, with, for example, the idea of 'slow tourism' (see Barr, 2018) emphasising the environmental advantages of substituting rail for air journeys when selecting and travelling to/from holiday destinations. John Preston and Simon Blainey commend rail's 'energy efficiency per traffic unit that other modes cannot match, with commensurate environmental benefits'. At the same time, they note that the mode's environmental advantage has been eroded as innovations elsewhere (e.g., electric cars) have outpaced developments in rail technology, and that delivering a truly net zero railway is not necessarily an easy proposition. Using a framework of 'nine grand narratives', they outline how this might be achieved through action by the rail industry, government, and end users adopting strategies to improve their current practices, shifting the emphasis from existing activities to other, greener, alternatives, and avoiding certain endeavours altogether.

Chapters 10 and 11 cover arguably the most difficult modes to decarbonise, aviation and shipping. Evangelia Pantelaki and Lucy Budd lament that claims 'for a future in which 'sustainable' aviation is possible have been made for over 20 years' in response to the sector having been pushed repeatedly by policymakers, politicians, and consumers. Unfortunately, 'contemporary evidence suggests there is still a long way to go before the promised solutions … mitigate aviation's environmental externality effects'. After discussing aviation emissions, Evangelia and Lucy provide a critical evaluation of the various technological propositions (that revolve around fuel and propulsion mechanisms) and regulatory interventions (ranging from banning certain domestic flights where feasible rail alternatives are available, to market-based approaches such as emissions trading schemes) that can be brought into play. They conclude that while the aviation sector has begun its journey towards net zero, 'the status quo cannot continue', and any realistic prospect of aviation decarbonisation will inevitably involve 'considerable financial investment, political will, technological innovation and concerted stakeholder engagement'.

The story is not entirely dissimilar in the maritime sector. Indeed, Jason Monios commences Chapter 11 with the warning that the sector 'is not yet serious about decarbonisation'. Like Evangelia and Lucy's, Jason's first substantive section covers the sector's energy demand and associated emissions. He then reviews potential alternative fuels and operational methods before evaluating the role of policy that, as with the aviation sector, involves a range of measures including emissions trading. In his conclusion, Jason sets out six clear policy steps that in his view need to be enacted as part of a strong, international policy response. He asks whether policymakers and the industry itself are ready to accept them but doesn't think the prospects are promising: 'Excuses are already in place and industry representatives expect the sector to be largely fossil-fuelled well past 2050, probably until around 2070'.

In Chapter 12, Jonathan Köhler, Clemens Brauer, and Lóránt Tavasszy focus on the decarbonisation and digitalisation transition in logistics systems, in the first of two contributions dealing specifically with transport's role in different economic sectors. The logistics industry is undergoing something of a transformation, with continuing growth of supply chains and demand for logistics services, not least in the context of burgeoning e-commerce activity. A key problem, however, is that while 'it is true that digitalisation ... increases transport efficiency ... [the increase] it is not expected to be sufficiently strong to allow the system to achieve its decarbonisation targets without intervention'. Such intervention requires policy modelling which addresses the processes of digitalisation and the interaction between digitalisation and decarbonisation. In their chapter, Jonathan, Clemens, and Lóránt develop a transition scenario that they assess through simulation modelling, concluding that 'socio-technical simulation could help stakeholders to understand which are the feasible and effective pathways towards change'.

Chapter 13 is concerned with tourism and travel. By 2030, tourism-related transport is expected to account for more than 5% of global emissions (UN World Tourism Organization, 2019), and Luca Zamparini notes there are questions about whether decarbonising the tourism sector is possible. This is because of obstacles including the relentless growth of the industry, the institutionalisation of interests, flawed policymaking, and the nature of the geography of tourism (people increasingly like to travel to destinations that they reach by air). In addressing CO_2 emissions produced by travel to and from tourist destinations, Luca makes the now familiar argument that technological developments alone will not be sufficient, and large-scale shift in behaviour and policy development – nothing less than a shift of paradigm – will be necessary. Even then, Luca reminds us that this is only part of the story. Accommodation and catering services in tourist destinations, not to mention the activities that tourists engage in, all have their own carbon emissions, a good proportion of which are transport-related thanks to fossil fuel-dependent supply chains. There are also tourist destinations themselves, for example cruise ships, that are transport based in their own right. Fortunately, efforts are underway to promote the kind of shared visions necessary to achieve net zero tourism, but no-one should pretend there will be anything other than a hard road ahead.

Chapters 14 and 15 discuss progress towards transport net zero through a geographical lens. Maria Attard in Chapter 14 looks at urban transport in net zero cities. Urban areas are home to over half of the world's population and in Europe generate 23% of all transport-related CO_2 emissions. The huge advantages European urban dwellers derive from their transport networks come at high cost, not just in terms of carbon emissions but also congestion, pollution, safety challenges, noise, and social exclusion. Maria explores the policy challenges of balancing benefits and costs through numerous examples in a discussion that unfolds over five sections: public transport, active travel, and micromobility; freight transport; multimodality; land use and transport integration; and the smart city and technological fixes. She also stresses the point that both technological advancement and behaviour change will be needed, and ultimately, her message is one of systems thinking:

> the future of city transport systems will depend on how well we manage to integrate the pieces and develop strategies that complement each other, rather than work in silos and designing systems and infrastructures that serve only one mode (the car) and are not inclusive.

In Chapter 15, Daniel Oviedo, Maria A. Wilches-Mogollon, and Natalia Villamizar Duarte examine transport net zero initiatives in rapidly growing economies, with a focus on Latin America. While some of the challenges, such as the dominance of the private car, will be familiar to policymakers in other countries, Latin America's 'urbanisation dynamics, informal transport systems and stark socio-spatial inequalities' present particular complexities. At the same time, they bring opportunities, and Daniel, Maria, and Natalia exemplify Latin America's history of pioneering sustainable transport initiatives. Active travel schemes involve a range of approaches from closing streets to promote cycle use to co-creating school programmes with children and adolescents. Bus rapid transit schemes have come to play 'a fundamental role in improving urban mobility in dense and congested cities'. Aerial cable cars have become an increasingly common part of the urban transport landscape across Latin America since 2004. Participatory and inclusive planning approaches are especially promising, not least because they offer a direct means of adapting to often diverse local circumstances in a situation where 'the lessons from the region underscore that achieving low-carbon mobility is inseparable from the broader goals of social inclusion and urban resilience in the global south'.

The last four chapters of this book address decarbonising the transport sector from different, but complementary, social and demographic perspectives. Lucy Budd focuses on gender and the transition to net zero transport in Chapter 16. Until recently, relatively little has been written relating to gender and the gendered implications of the move towards net zero transport, which is perhaps unsurprising given that gender equality, transport, and climate change have long been seen as separate from each other. It is also extremely unfortunate. Between the genders, there are differences in travel and employment experiences, environmental attitudes and behaviours, and average carbon intensity of mobility. But as Glenn Lyons (2004) wrote over 20 years ago, transport systems seem to have been shaped around the needs of men, and particularly middle-class, middle-aged professional ones at that. Lucy first constructs a gender perspective in the context of decarbonising transport, before moving on to discuss the nature of women's work in the transport sector. Not only are women under-represented in the workforce (constituting around 17% globally) but they also face particular challenges at work and as users of transport systems – including physical and sexual assault – because of their gender. Lucy concludes that transport decarbonisation strategies need to include a gender perspective and address issues of diversity in the workforce, not least because 'a transport system that is not designed for and with everyone (irrespective of age, gender, race, nationality, or disability) will not achieve its decarbonisation targets'.

In Chapter 17, Charles Musselwhite discusses ageing, transport, and decarbonisation. The current cohort of older people has in many countries driven more than any other generation, and although studies point to the fact that they can find it very difficult to stop driving, Charles suggests that older people shouldn't be absolved from the responsibility to use their cars less on the basis of age alone. It may well be that targeted approaches to help older people engage in more sustainable mobility behaviours are necessary, and there is certainly evidence to suggest that members of the cohort can feel guilty for their contribution to climate change and want to be an active part of measures designed to address the problem. This finding is significant because the mobility experiences of older people in their much younger days – more walking and cycling before cars became increasingly ubiquitous, and much less flying – offer practical ideas for younger people looking to change their own mobility practices. Charles unpacks these issues before setting out different kinds of policy interventions that, properly delivered, can make alternatives to car-dependent lifestyles achievable for older people or at least reduce their reliance on private cars.

In Chapter 18, Claus Hedegaard Sørensen, Tom Rye, and Lisa Hansson confront head on the need to ensure the political legitimacy of transport strategies and policies necessary for net zero to be achieved. Their focus is on one such means of doing this, namely, public participation. Taking as their starting point 'much experience that illustrates the difficulties of gaining legitimacy if the population is not involved in the [policy making] process', Claus, Tom, and Lisa are under no illusion that public participation is easy; getting it wrong can cause more problems than it solves. In order to maximise the chances of success, authorities need to understand: who it is they are trying to seek participation from; how to make people aware their participation is being sought; whether people possess the kind of knowledge relevant to the questions being asked of them; and the extent to which consensus can ever be an ideal (or indeed realistic) goal. Elaborating on benefits of participatory planning identified in Chapter 15, the authors advance the notion of representative and deliberative participation as a means of trying to ensure the legitimacy of process outcomes. While they concede there is no silver bullet, for them a transformative net zero policy agenda is more likely to yield better results with a credible participation exercise than without one.

Finally, in Chapter 19, Corinne Mulley and Maria Attard – names now well-known to readers of this volume and indeed book series – reflect upon the disruptive impact of COVID-19 on transport systems and mobility. They do this to highlight governance and behaviour change lessons that could inform climate change mitigation and adaptation efforts. There is potentially fertile ground here, and 'the ideal of maintaining and delivering growth whilst also reducing consumption and limiting emissions is possible through the lessons learnt from the COVID-19 pandemic recovery'. As things have turned out, subsequent events 'attracted global attention away from the climate challenge [and] there is an urgency to look back and understand the potential of growth through more sustainable mobility'. Overcoming this will require governance processes to be formulated in the same 'spirit of global cooperation' as during the pandemic, but the completely different temporal nature of the threat – climate change has been happening far more slowly than the sudden, sharp shock of lockdowns – is bound to militate against such an outcome in the context of current geopolitical challenges among other things. Their concluding sentences are salutary:

> Ultimately the issues we are facing due to climate change are not new but are ones that could lead to outcomes more severe than the COVID-19 pandemic. Unlike the pandemic, however, there is no vaccine that would see a slow but sure recovery for the planet and its inhabitants.

TOWARDS TRANSPORT NET ZERO? SOME CONCLUDING REMARKS

As the last sentence suggests, while achieving net zero may be an enormous challenge, it is a necessary one to overcome. Doing this 'calls for nothing less than a complete transformation of how we produce, consume, and move about' (UN, n.d.), and delays or inaction will only serve to make the transition more difficult and costly for current and future generations. In this volume, we provide evidence in support of urgently needed action among governments and the communities they serve.[3] We collate and present the findings of research into gaps that persist in understanding and implementing the promise of technology and in identifying the most effective ways and means of changing people's existing travel behaviour. What are the key themes and potential future directions that emerge?

Perhaps before anything else, we need to be clear about what 'transport net zero' actually means; a gap in understanding around this fundamental concept muddies the waters for everything that follows. To us as editors, transport net zero is the position where CO_2eq emissions from the transport sector – from going to the local shop for a litre of milk, to the supply chains that deliver it; from a Sunday afternoon trip to the park to a two-week summer holiday abroad – are sufficiently reduced that they 'can be absorbed and durably stored by nature and other carbon dioxide removal measures, leaving zero in the atmosphere' (UN, n.d.). Since the sector is responsible for around a quarter of global greenhouse gas emissions, the scale of the challenge is truly enormous. It behoves us to remember that this is not some abstract, future inconvenience: as we write Storm Éowyn is wreaking havoc in much of the United Kingdom and the Republic of Ireland (where the strongest ever wind gust, of 183 km/h, has just been recorded). Changing climatic conditions today – literally and figuratively – bring to the fore the need to adapt our systems to deal with their current vulnerabilities as well as mitigate by reducing their emissions in the years to come.

The good news is there has been progress across a number of fronts, at various scales, in all manner of places around the world. Less positively, it has tended to be patchy and non-linear: witness Donald Trump's Executive Orders hours after his inauguration reversing measures designed to promote electric vehicles and withdrawing the United States from the Paris Climate Agreement for the second time. Worryingly there remains an evident (and industry-driven) emphasis on technological solutions that primarily serve the global north. The UN Secretary-General's recent 'code red for humanity' statement (Gutierrez, 2021) should have been a stark wake-up call for us to realise that, even with all the investments in vehicle technologies made in the last 50 years, transport unlike other sectors has still not managed to bring down its overall emissions. There is no escaping that too little has been spent on looking at ways in which technologies will be implemented, used, and transferred. Behavioural shift can curb emissions by reducing relevant consumption, but it can be difficult to achieve. We would argue for more grounded, applied, and action research that is co-created and participatory, and thus embedded in communities to whom the benefits of change should most tangibly flow. Emerging policy imperatives on transport poverty (see, for example, the European Union's (EU) latest funding programme) will certainly provide a new emphasis on the need for a socially just transition to net zero. But research gaps exist between populations, and policy development and implementation differ spatially as different parts of the world continue to 'drive' economic growth.

What is also clear is that single, fragmented policy interventions are less effective than a package of complementary measures and/or approaches that stress the 'co-benefits' – i.e., to a range of constituencies – of actions to address a specific issue (see Birner et al., 2024; Karapin & Vogel, 2023). For example, advocates of congestion pricing struggle to sustain momentum for schemes that are not framed within a wider strategy that seeks to provide for those who cannot walk or cycle and for those who would be excluded by the high cost or limited availability of public transport alternatives. Successful strategies also require interdisciplinary approaches to understand their full range of costs and benefits. It is no longer sufficient to champion projects with narrowly defined transport or economic objectives. Successful investments will be appraised (ex ante) and evaluated (ex post) on the basis of the extent to which they also take into consideration environmental impact, resilience, social justice, and public health and well-being. As we say, progress has been made, and while best practice in one location is by no means guaranteed to work everywhere else, there is much to be gained from institutional and social processes of policy transfer and mobility (Cochrane & Ward, 2012; Dolowitz & Marsh, 1996, 2000; Ison et al., 2011) supported by methodologies, case studies, and other experiences of successful policy implementation that can be adapted to suit local circumstances.

A full and meaningful transport sector contribution to broader sustainable development is not optional. Achieving net zero is one part of such a contribution, and as we and the contributors have been at pains to stress, this is not solely about inventories and numbers. Sustainability in transport is to attain economically productive, socially just, green movement and mobility. The UN's Sustainable Development Goals (UN, n.d.) establish a framework that goes beyond net zero, and in its pursuit, we cannot lose sight of what this means for our transport systems that empower and enrich our lives and in so doing provide opportunities of access to education, jobs, goods, healthcare, and leisure. As editors of the *Transport and Sustainability* series, we hope the volumes already published have made worthwhile individual and collective contributions in this regard. It is our aim to continue providing a platform for researchers to showcase the ideas derived from their research about how best we shape a genuinely sustainable transport system for generations to come.

NOTES

1. Carbon dioxide equivalent (CO_2eq) is a way of expressing the global warming potential of different greenhouse gas emissions in terms of the global warming potential of a metric tonne of CO_2 emissions. To illustrate, methane has a global warming potential 25 times greater than CO_2, hence 1 tonne of methane is 25 tonnes CO_2eq (Eurostat, n.d.).
2. Electric and AV technologies sit very well together and as such are ripe for co-development. In his book *Kaput*, Wolfgang Münchau (2024, pp. 61, 71) explains this

2024 were at 96% of pre-pandemic levels (UN Tourism, 2024). Business trips have been slower to recover, but a mid-2024 survey of US-based companies reveals that spend on corporate travel is expected to reach and perhaps surpass 2019 levels by the end of 2024 (Deloitte, 2024). This is despite the impact of COVID-19 with business meetings conducted online and increased working from home. The situation is stabilising post-COVID-19 at around one or two days of home working for many workers in the Western world, confirmed in existing literature and a survey of Sydney-based employees (Hensher et al., 2024).

The majority of the transport system is reliant on fossil fuels, with oil products used for nearly 91% of its final energy (International Energy Agency (IEA), 2023). Most governments and many non-state actors, such as companies and educational institutions, have committed to being net zero by 2050 (United Nations, 2024a). Addressing the CO_2 emissions of transport activities is a major component of this commitment. In progress towards a more sustainable transport system, there are requirements for a reduction in travel by modes with high CO_2 emissions.

With most transport emissions coming from the road sub-sector (75% of global transport CO_2 emissions in 2023, according to Statistica (2025)), the challenge to reduce automobile dependence in cities is even more relevant now than when it was originally demonstrated 30 years ago (Newman & Kenworthy, 2021). Over mid-to-long distances, there are transport modes with limited sustainable alternatives, namely aviation and shipping, which have substantially increased in use over time. Transport emitted 8 Gt CO_2 globally in 2016, 71% more than 1990 levels, but it is sea (84%) and air (115%) which had the greatest proportional increase (IEA, 2018). One recent study by Wadud et al. (2024) used English data to demonstrate the impact of long-distance travel. They calculated that even though just 2.7% of a person's trips are for long-distance travel (>50 miles one way), they account for 61.3% of the miles and 69.3% of the greenhouse gas (CO_2 equivalent) emissions.

The situation is complex given that the climate change impact of transport activities is dependent on a range of factors, including the emissions levels and load factors of vehicles, aircraft, and ships. An international assessment of progress towards a new zero emission pathway has outlined that net zero means getting as close to no greenhouse gas emissions as possible, with any remaining emissions re-absorbed from the atmosphere, for instance, by oceans and forests (United Nations, 2024a). By 2050, of the six transport categories reviewed, only one of them, electric vehicles, has a realistic chance of reaching net zero (IEA, 2023; see Chapter 5). The remaining five categories either need improvement or are not on track: cars and vans, trucks and buses, rail, aviation, and international shipping (see Chapters 9–11).

Climate change is certainly a major global challenge. As documented by the Intergovernmental Panel on Climate Change (IPCC, 2023), there is unequivocal evidence that human activity is already changing the world in unprecedented and sometimes irreversible ways. The Sixth Assessment Report (IPCC, 2023) stated that global surface temperatures reached 1.1°C above 1,850–1,900 levels in the decade 2011–2020. The 2021 data within the report highlighted that warming will likely exceed 1.5°C during the 21st century; this is important since it is moving further along the trajectory to what is regarded dangerous warming of 2°C. More recent data from the World Meteorological Organization (WMO, 2023) confirmed not only that 2023 was the warmest year on record, but also that global average near-surface temperature has already reached 1.45°C (with a margin of uncertainty of ±0.12°C) above the pre-industrial baseline (WMO, 2023).

The difference between mitigation and adaptation remains a useful framework for examining the relationship between transport and climate change, as put forward in the 2001 IPCC Synthesis Report (IPCC, 2001). Given the current level of emissions, there remains an ongoing need to mitigate the emissions which cause climate change. This is compounded by the forecast that the growth in transport is set to continue. There is an expected major increase up to 2050 (from 2015 figures) across both personal (2.3-fold) and freight (2.6-fold) transport (International Transport Forum (ITF), 2021). Much of the mitigation focus is on developing alternative fuels to power transport, including electric, biofuels, and hydrogen as possible sole or hybrid sources. Technological advances are all the more significant in the context that attempts to promote behaviour change remain politically very difficult. Most individuals would find it difficult to live without their car and alternatives struggle to overcome this challenge, as demonstrated by various travel behaviour surveys (e.g., Alyavina et al., 2024; Lucas & Jones, 2009). Similar results have emerged from air travel behaviour surveys, with very few individuals willing to fly less (e.g., De Vos et al., 2024; Ryley & Davison, 2008). Unfortunately, no single measure, whether technological or behavioural, provides a solution to stabilise and reduce transport emissions; multiple interventions across the spectrum from 'hard' technological (e.g., electric vehicles – see Chapter 5) to 'soft' behaviour change (e.g., reducing car use in favour of the benign modes of walking and cycling – see Chapters 3 and 4) will be required.

At the same time, with the rise in extreme weather events and associated impacts of climate change, such as increased air and water temperatures, flooding, and severe winds, there is a growing need for climate change adaptation by the transport industry. IPCC (2022) provides an outline of the weather scenarios across different regions of the world. One example provided, for Australasia, will mean more hot days, reduced snow cover, ongoing sea-level rise, ocean acidification, and fewer but more severe tropical cyclones. Issues split by the mitigation and adaptation framework, albeit framed in 2012, highlighted the mitigation aspects of car ownership, freight and aviation, and adaptation required around climate change impacts and risk assessment (Chapman & Ryley, 2012).

It is therefore clear that there are challenges of rising demand for leisure and business travel, continued over-reliance on fossil fuels with limited sustainable alternatives, and the scale of road transport emissions and long-distance travel, typically by air and sea. This is alongside worsening climate change conditions, such as increased global temperatures and more extreme weather events. In this chapter, these realities have been reviewed and re-framed as a series of five major issues relating to transport and climate change which need to be addressed. First, climate change is a global problem which transport actors, and organisations need to resolve across the spatial scales. Second, with a technological emphasis, there is a need to accelerate and scale up innovative solutions. Third, it is evident that difficult climate change and transport decisions need to be taken by industry and individual travellers. Fourth, with an increase in extreme weather events, there needs to greater climate change and transport adaptation plans. Fifth, as the impact of climate change and associated weather events become more extreme, so aspects of injustice will require attention. In discussing these five issues, this chapter is not intended to be comprehensive but rather to unpack key points that demonstrate the importance of progressing a net zero transport agenda.

FIVE MAJOR ISSUES IN TRANSPORT AND CLIMATE CHANGE

Climate Change Is a Global Problem That Needs to Be Addressed Across the Spatial Scales

There are acute spatial challenges for less-sustainable long-distance travel, namely by air and sea, where national and international governance responsibilities and associated emissions accountability can be complex and blurred. As identified in a literature review of aviation adaptation (Ryley et al., 2020), air transport is locally situated in airports, represents regional interests through national airlines, and crosses geopolitical boundaries in its operations. These extensive networks of responsibility and interest involve all levels of governance from city planning to the International Civil Aviation Organization (ICAO), and a range of actors from individual passengers to multinational corporations and institutions. Therefore, it is hard to assign aviation emissions, with complexities across airports flown to-and-from, passenger nationalities, and airspace traversed. As Budd and Ryley (2012) note, the flight path taken can have an impact on carbon emissions, as well as flight time length and airline operating cost. This could be the prescribed flight plan, as well as deviations from the filed plan, say due to storms en route. Prioritising carbon emissions against flight costs is a challenge, as is the generally spatial challenge of finding an appropriate route navigating meteorological and geopolitical situations.

Climate change responses, including for the transport sector, tend to focus on international agreement and action, principally stemming from the annual Conference of the Parties (COP) to the UN Framework Convention on Climate Change. The Paris Agreement, from COP 25 in December 2015, resulted in 196 nations signing a binding treaty to hold global average temperatures to well below 2°C compared to pre-industrial levels, and the attempt to limit rises to 1.5°C. COP 28 in the UAE in December 2023 led to an agreement to signal the beginning of the end of fossil fuels (United Nations, 2024b), although attempts to keep temperatures below a 1.5°C increase are unlikely to be successful.

The merits or otherwise of the COP have been discussed recently in November 2024 at COP 29, in Baku, Azerbaijan. The COP process brings world leaders together to debate climate change issues and deliver on a series of agreements and financial support packages, but there is underlying division between countries, for example, between the developing world who argue for greater change and financial assistance and those from wealthier countries with a greater per capita contribution to climate change (*The Guardian*, 2024a). Holding the COP 29 in Azerbaijan also came under criticism as the host country is over-reliant on fossil fuels and arguably lacks a desire for real change on decarbonisation action (Besliu, 2024).

Yet while climate change at the headline level is a global challenge, it has implications at continental, national, regional, and local scales. A key problem is that top-down, international policy and management frameworks can be difficult to implement at regional and local levels. Some commentators, such as Dirix et al. (2013), stress the need to instigate a bottom-up approach, so as to build domestic support for future climate treaties, empower citizens, and motivate leaders to take action. Perhaps, as suggested by Chapman and Ryley (2012), it would be sensible to also think about local transport mitigation measures before subsequently scaling up to national and international levels.

The slow progress of climate change policy implementation has led, in part, to the growth of bottom-up protest, as typified by the *flyskam*, or flight shaming, movement which emerged in Sweden in 2018 and was popularised by the environmental activist Greta Thunberg. Thunberg has been among the critics of the COP process, labelling the most recent COP 29 as greenwashing (*The Guardian*, 2024b). Transport leaders and organisations, particularly in areas with more pro-environmental attitudes such as Northern Europe, need to incorporate into their policies and practice both the global policy framework and the energies of motivated individuals advocating change.

Transport policy across multiple governance levels makes delivery complex and fragmented (see Marsden et al., 2014). As Urwin and Jordon (2008) state, after examining top-down and bottom-up approaches, neither one offers a complete picture of the potentially enabling or constraining effects of different policies on future adaptive planning. Together they offer new perspectives on climate, and the hope is for meeting in the middle between policymakers and individual travellers. In many ways, cities are a middle-level challenge and an opportunity for climate change policy (Corfee-Morlot et al., 2009). As the hubs of economic activity, cities generate the bulk of greenhouse gas emissions and are thus important to mitigation strategies (see Chapter 14). Urban planning will shape future trends and the concentration of population.

Certainly, transport actors and organisations need to be agile and prepared in response to such spatial challenges. Integration is key, as there is a politically difficult but potentially powerful combination in harnessing bottom-up initiatives to help achieve top-down goals.

Transport Technological Innovations Which Reduce Carbon Emissions Need to Be Accelerated and Scaled Up

Various transport technological innovations are under development to reduce carbon emissions. Much of the mitigation focus has been on fuel development, typically across electric, hydrogen, and biofuel alternatives, including hybrid options. Electric vehicles are a relative success story and on track for net zero by

The role and impact of carbon offsets needs to be clarified, as there are questions relating to them as a viable solution to remove equivalent emissions. Offsetting works on the principle of carbon neutrality by paying another actor to compensate for the emissions generated, but the scope, quality, and efficacy of offset schemes vary dramatically. This is certainly the case for air travel. Airline information on the offset options, such as tree planting and investments in aircraft technologies, is inconsistent, limited, and sometimes inaccurate (Becken & Mackey, 2017). Furthermore, as Work for Climate (2022) notes, carbon neutral strategies can be unregulated, insufficiently scientifically rigorous, and overlook Scope 3 emissions, and in any event do not necessarily incentivise behaviour that reduces carbon emissions in the first place. A net zero approach on the other hand requires much more stringent, sector-wide action to slash absolute emissions to a point where offsetting is credible at a global scale (Work for Climate reports this at around 10% of current emissions).

There are clearly difficult and unpopular transport-related climate change decisions required by government, industry, and individual travellers, who will face rising costs associated with climate change. Many transport organisations need to deepen their understanding of the emissions they are responsible for and develop appropriate responses accordingly. Issues of climate change are becoming increasingly important in relation to finance and investment, particularly when some notable transport organisations such as the airline KLM have recently been taken to court over greenwashing claims (Reuters, 2024).

The Transport Industry Needs to Be Planned and Ready for Climate Change Adaptation

The underpinning data analysis of the IPCC Sixth Report (IPCC, 2023) has generated a high level of confidence that there have been widespread extreme adverse weather impacts, such as from heatwaves, heavy precipitation, droughts, and tropical cyclones, with related losses and damage to nature and people. Climate change impacts on the transportation system can be from direct and sudden weather events or associated long-term knock-on effects such as sea-level rise. While many people can face major travel disruption as a result, some individuals have become increasingly resilient, choosing not to travel and working from home in response to challenging weather situations, as evidenced by examining weather data against broadband usage (Budnitz et al., 2023).

There are challenging implications for transport systems though, such as the road, rail, and aviation networks. In the United Kingdom, as one national example, there has frequently been major disruption to both the national road and rail networks, such as the flooding that affected one of the primary motorways, the M5, in October 2024 (BBC, 2024). In an extreme example, flooding in the summer of 2007 led to many flooded roads including the M5, on which 10,000 motorists in Gloucestershire were stranded overnight (Royal Meteorological Society, 2008). Flooding in Dawlish, Devon, in 2014 led to the closure of the main railway line in south-west England for eight weeks (Dawson et al., 2016; Network Rail, 2024).

There is a wide range of impacts of climate change on air travel operations, both in flight and at airports. A recent literature review (Holmes et al., 2024) identified flight operation climate change hazards to include changes in wind, turbulence, dust, smoke, icing, and hail and airport operation hazards including changes in precipitation, heat, saltwater inundation, tsunamis, lightning, and volcanic ash. As shown by an aviation adaptation review (Ryley et al., 2020), modelling the weather impacts must incorporate associated complications, such as jet streams and clean-air turbulence, as well as cover a range of outcomes from aviation safety through to cost implications. Adaptation examples in aviation include the industry responding to demand changes from tourism trends, low-lying airports affected by rising sea levels, and flights re-routed due to weather delays (Budd & Ryley, 2012).

In April 2024, one of the busiest airports in the world, Dubai, faced major disruption due to flooding (Australian Financial Review, 2024). Such events cause cancellations, rerouting, and diversions for travel, plus major organisational and scheduling challenges for the transport companies involved, as well as major financial implications. As a related example, Southwest Airlines faced severe financial penalties due to bad weather forcing flight cancellations in Winter 2023 that left it struggling to resume operations and appropriately compensate passengers (*The New York Times*, 2023; US Department of Transportation, 2023).

A review by Mattsson and Jenelius (2015) looked at transport disruption in terms of the reliability and vulnerability of a system. Weather and climate impacts were seen as having the greatest impact (some such as natural disasters may not be attributable to climate change), along with terrorism incidents, but transport organisations must also account for more minor events associated with traffic accidents, engineering works, technical failures, and civil disputes. Despite the complexity involved and major impacts, Marsden et al. (2020) argue that there is potentially a much greater adaptive capacity in response to disruptions in the mobility system than is often given credit. Data suggest that this could happen due to a rapid increase in innovation, flexibility in transport service provision (e.g., with mobility as a service (MaaS) – see Chapter 8), and societal changes associated with the way people live, work, and travel (see Chapter 19).

With the increasingly complex and varied weather and climate impacts on the transport system, the transport industry needs to be planned and ready for climate change impacts, but there are concerns about the extent that it is. Transport vulnerability and impact analysis requires linking to future climate projections to allow assessment of the probability and frequency of future climate change-related events on networks (see Dawson et al., 2016). Good practice sustainability plans, as outlined when they emerged for large international airports, include identification of the major risks as well as assessments of the occurrence probabilities and action plan implementation (Thompson, 2016).

The Transport Sector Needs to Consider the Climate Injustice Aspects

Global differences in development have meant that vulnerable communities who have historically contributed the least to current climate change are disproportionately affected (IPCC, 2023). As stated in the discussion around the COP process, there is an increasing division between the developing countries and the developed world. The underlying data highlight these divisions. Table 2.1 shows car ownership rates for the largest countries in each of the four main global regions. There is a clear distinction between developed and developing countries, for example, comparing the relative high car ownership rates between Europe and the Americas with those in the largest African countries shown in the table. The overall variation between countries is from the lowest ownership rate of 4 cars per 1,000 inhabitants of Somalia right up to 1,608 cars per 1,000 in San Marino. The disparity also shows the growth potential of some of the developing countries, and associated emissions, if left unchecked.

Even within developed countries, there are significant anomalies. Various studies have demonstrated that a select few contribute to most transport-related emissions (Gössling & Humpe, 2020; Wadud et al., 2022). Wadud et al. (2022), using English data, showed that the top 5% of car users travel nearly 27,500 miles (4.8 times the average) and emit 5,742 kg CO_2 annually (5.7 times the average), and that those more likely to record higher mileage were male, high-income professionals, and owners of large cars. The stark difference also exists for air travel in

Table 2.1. Car Ownership Rates per 1,000 Inhabitants, Split by the Largest Five Countries (by Population) in Each of the Main Global Regions.

Country	Population (Millions)	Car Ownership Rate
Asia		
India	1,450	158
China	1,419	210
Indonesia	283	485
Pakistan	241	83
Bangladesh	174	18
Africa		
Nigeria	233	60
Ethiopia	132	7
Egypt	117	83
Tanzania	69	40
Kenya	56	62
Americas		
United States	345	863
Brazil	212	462
Mexico	131	332
Columbia	53	280
Argentina	46	493
Europe		
Russia	145	371
Turkey	87	258
Germany	85	684
UK	69	583
France	67	653

Sources: Our World in Data (n.d.); Worldometer (2024).

Note: Our World in Data reported the total number of registered vehicles (i.e., vehicles reported to a government agency and given some form of registration) in 2017 per 1,000 people in each country. There was no car ownership data for DR Congo (population 109 million) or South Africa (population 64 million), so they are omitted here.

England, although could be replicated in other developed world countries. Using data from a Department for Transport survey, *The Guardian* (2019) reported that 1% of English residents took around 20% of all flights abroad in 2018 and 10% of the most frequent flyers took over half of all international flights. O'Garra and Fouquet (2022) conclude that behaviour associated with affluence represents a major barrier to a low-carbon transition, and that policies must address over-consumption by those with the greatest spending power as a priority.

It is not just wealth that is a major issue in relation to climate inequality and injustice but also age, gender, and other characteristics. A wider perspective is therefore required by the transport sector when such questions are addressed, not only to consider different world views but also to appreciate a deeper understanding of theory, methods, and research on climate change and transport – particularly beyond the oft-used mitigation and adaptation dichotomy, as discussed in this chapter – and further explore concepts such as complexity and uncertainty (Schwanen, 2019). What is certain is that this brief discussion only scratches the surface relating to climate change injustice associated with the transport sector. There is a major affluence disparity both between countries and within nations in terms of transport usage and carbon emissions. There is also wider injustice across other dimensions which requires a greater focus and response from the transport sector (see Chapters 16 and 17 for discussions of gender and age in relation to transport net zero).

CONCLUSION

This chapter has sought to underline the importance of the towards transport net zero agenda by focusing on five major issues, related to spatial complexity, technological innovations, decision-making challenges, the need for adaptation, and climate injustice. Together they provide a useful framework for future research relating to transport and climate change. Transport actors and organisations need to be agile and prepared in response to spatial challenges. Technological innovations need to accelerate and scale up alternative fuels, as well as increase the penalty for using fossils fuels, say through existing taxation and incentive mechanisms. There are difficult demand management decisions to be made, including the need to avoid and reduce the amount of transport undertaken. There needs to be a stronger focus on adaptation as well as urgency in mitigation. And progress towards transport net zero has to be just.

The transport sector still has longer-term net zero aspirations. However, given that recent global temperature increases are ahead of predictions, it needs to urgently accelerate progress. The three primary conclusions of previous work on transport and climate change (Chapman & Ryley, 2012), more than a decade ago, seem as pertinent now as they did then. First, early responses are crucial to minimise climate change costs; it could be that society is living on borrowed time to avoid 'dangerous' climate change levels. Second, there is a need to remain focused on the climate issues at hand when other, short-term, crises emerge; there are concerns that, for example, economic downturns after the Global Financial

Qantas. (2024). *Sustainable aviation fuel*. Retrieved January 28, 2025, from https://www.qantas.com/au/en/qantas-group/sustainability/our-planet/sustainable-aviation-fuel.html

Reuters. (2024, March 21). *Dutch court finds KLM ads were misleading in 'greenwashing' case*. Retrieved January 28, 2025, from https://www.reuters.com/business/aerospace-defense/dutch-court-rule-klm-greenwashing-case-2024-03-20/

Royal Meteorological Society. (2008). *Editorial: Special issue – The wet summer of 2007 in the UK*. Retrieved January 28, 2025, from https://rmets.onlinelibrary.wiley.com/doi/pdf/10.1002/wea.323

Ryley, T., Baumeister, S., & Coulter, L. (2020). Climate change adaptation for the aviation industry: A systematic quantitative literature review. *Transport Policy*, *92*, 54–64.

Ryley, T., & Davison, L. (2008). UK air travel preferences: Evidence from an East Midlands household survey. *Journal of Air Transport Management*, *14*(1), 43–46.

Schwanen, T. (2019). Transport geography, climate change and space: Opportunity for new thinking. *Journal of Transport Geography*, *81*, 102530.

Statistica. (2025). *Distribution of carbon dioxide emissions produced by the transportation sector worldwide in 2023, by sub sector*. Retrieved January 28, 2025, from https://www.statista.com/statistics/1185535/transport-carbon-dioxide-emissions-breakdown/

Stradling, S. G., Meadows, M. L., & Beatty, S. (2000). Helping drivers out of their cars. Integrating transport policy and social psychology for sustainable change. *Transport Policy*, *7*, 207–215.

The Guardian. (2019). 1% of English residents take one-fifth of overseas flights, survey shows. *The Guardian*, September 25. Retrieved January 28, 2025, from https://www.theguardian.com/environment/2019/sep/25/1-of-english-residents-take-one-fifth-of-overseas-flights-survey-shows

The Guardian. (2024a). Developing countries urged to reject 'bad deal' as Cop29 climate talks falter. *The Guardian*, November 23. Retrieved January 28, 2025, from https://www.theguardian.com/environment/2024/nov/22/developing-countries-told-to-reject-bad-deal-as-cop29-talks-falter

The Guardian. (2024b). A 'Cop of peace'? How can authoritarian, human rights-trashing Azerbaijan possibly host that? *The Guardian*, November 11. Retrieved January 28, 2025, from https://www.theguardian.com/commentisfree/2024/nov/11/greta-thunberg-cop29-authoritarian-human-rights-azerbaijan-greenwashing

The New York Times. (2023). Southwest says holiday meltdown will cost it more than $1 billion. *The New York Times*, January 26. Retrieved January 28, 2025, from https://www.nytimes.com/2023/01/26/business/southwest-airlines-fourth-quarter-earnings.html

Thompson, T. R. (2016). Climate impacts upon the commercial air transport industry: An overview. *Carbon and Climate Law Review*, *10*(2), 105–112.

United Nations. (2024a). *For a liveable climate: Net-zero commitments must be backed by credible action*. Retrieved January 28, 2025, from https://www.un.org/en/climatechange/net-zero-coalition

United Nations. (2024b). *COP28 agreement signals "beginning of the end" of the fossil fuel era*. Retrieved September 16, 2024, from https://unfccc.int/news/cop28-agreement-signals-beginning-of-the-end-of-the-fossil-fuel-era

UN Tourism. (2024). *UN Tourism barometer|global tourism statistics*. UN Tourism.

Urwin, K., & Jordan, A. (2008). Does public policy support or undermine climate change adaptation? Exploring policy interplay across different scales of governance. *Global Environmental Change*, *18*, 180–191.

US Department of Transportation. (2023). *DOT penalizes southwest airlines $140 million for 2022 holiday meltdown*. Retrieved January 28, 2025, from https://www.transportation.gov/briefing-room/dot-penalizes-southwest-airlines-140-million-2022-holiday-meltdown

Wadud, Z., Adeel, M., & Anable, J. (2024). Understanding the large role of long-distance travel in carbon emissions from passenger travel. *Nature Energy*, *9*, 1129–1138.

Wadud, Z., Adeel, M., Anable, J., & Lucas K. (2022). A disaggregate analysis of 'excess' car travel and its role in decarbonisation. *Transportation Research Part D: Transport and Environment*, *109*, 103377.

Work for Climate. (2022). *Carbon neutral, net zero and zero emissions: What's the difference?* Retrieved January 28, 2025, from https://www.workforclimate.org/post/whats-the-difference-between-carbon-neutral-net-zero-and-zero-emissions

World Meteorological Organization. (2023). *Stated of the global climate 2023*. Retrieved January 28, 2025, from wmo.int

Worldometer. (2024). *Countries in the world by population (2025)*. Retrieved January 28, 2025, from https://www.worldometers.info/world-population/population-by-country/

CHAPTER 3

WALKING FOR NET ZERO

Yuting Zhang and Corinne Mulley

The University of Sydney, Australia

ABSTRACT

Walking is the most carbon-friendly form of transportation, releasing little carbon dioxide into the atmosphere. As an active mode, it is central to plans to offer mode shift away from the private car, lessening dependence on fossil fuels and bringing other environmental and personal health benefits. This chapter discusses these ideas and considers how walking might be measured, before turning to focus on planning and policy implications of building a robust walking infrastructure, in particular the need to improve the physical infrastructure and promote a culture of walking. Emphasising the importance of a holistic approach to sustainable mobility, the final section presents several ideas that have emerged across the globe for effective interventions and actions that encourage walking and illustrates the mode's future role in the net-zero agenda as a crucial element of climate change mitigation.

Keywords: Walking; net-zero agenda; planning policies for walking; physical activity and walking interventions; sustainability; climate change

INTRODUCTION

As a mode of transport, walking is unique. It is the most carbon-friendly form of transportation, releasing little carbon dioxide into the atmosphere. As a contribution to the climate change mitigation agenda, walking as an active mode is central to plans to achieve mode shift away from the private car, lessening dependence on fossil fuels and bringing other personal health benefits and providing a better environment in which biodiversity can thrive.

Towards Transport Net Zero
Transport and Sustainability, Volume 20, 27–43

ISSN: 2044-9941/doi:10.1108/S2044-994120250000020003

This chapter discusses the role of walking in helping to achieve net zero in the transport sector. The next section highlights the ever more pressing need to encourage a walking culture to ameliorate the lack of physical activity among the global population. How walking might be measured is also considered. Attention then turns to the policy and planning implications associated with promoting walking as a mode of transport. In building and encouraging a robust walking infrastructure, policymakers need to grapple not only with providing a high-quality physical environment conducive to walking – taking into account the location of infrastructure and the timing of its provision if walking is to be made safe and enjoyable – but also with nurturing a robust walking culture. The final section of this chapter highlights policies enacted in various parts of the world that show how walking can be used as a way of contributing to the net-zero agenda.

WALKING FOR A SUSTAINABLE FUTURE

The prevalence of obesity as a world-wide public health problem is highlighted by the World Health Organization (WHO) (2024), which estimates 16% of all adults over 18 were obese in 2022; the adult rate of obesity had doubled since 1990 and the adolescent rate quadrupled. The WHO first developed a web-based Health Economic Assessment Tool (HEAT) in 2005, and this has had a number of updates following new scientific evidence. The current version (5.0) of the tool looks to assess the health and economic benefits of increased walking and cycling. It examines how these activities can reduce the likelihood of premature death by accounting for changes in physical activity, exposure to air pollution, and the risk of fatal accidents, while also considering the impact of reduced carbon emissions from shifting away from motorised travel.

But the HEAT approach is not the only method of evaluation in existence, with national jurisdictions having implemented their own method and tools. In England, for example, the Department for Transport's WebTag library gives guidance on the estimation and reporting of impacts for active modes (Department for Transport, 2022), and the New Zealand Transport Agency introduced a revised *Monetised Benefits and Costs* manual in 2023 which has a separate section on the valuation of active travel (Waka Kotahi NZ Transport Agency, 2023). However, while the development of such methodologies has led to a more rigorous assessment of proposed projects, it remains the case that significant cost elements are often overlooked with the effect of favouring investment in the non-active modes. In a road scheme, for example, these overlooked elements might include generated traffic and downstream congestion, parking costs, vehicle ownership costs, indirect environmental costs, strategic development costs, increased accessibility for non-drivers, social equity considerations, and public health changes (Litman, 2017).

Litman (2017) identifies benefits and costs of walking arising from transportation systems as falling into four categories, each of which needs to be appraised to identify a consistent valuation: improved walking conditions, increased walking activity, reduced private car travel, and more compact communities. Some of these elements are well documented in research. Grid-like built environments are associated with more walking (Aliyas, 2019; Koschinsky et al., 2017; Kowaleski-Jones et al., 2018). More recently, there has been attention paid to place making – the process of designing and enhancing public spaces to promote community interaction and improve quality of life – and its role in creating a more walkable environment (Carmona, 2019). Increased walking activity also provides benefits in terms of user enjoyment and better public health, and Loukaitou-Sideris (2020) notes that more walking activity increases social interaction within neighbourhoods which in turn leads to greater security. The most overlooked category of change relates to the way in which walking improvements can lead to more compact neighbourhoods which tend to have lower private car use (although this is not guaranteed – see Stevens, 2017), greater social cohesion and higher satisfaction (Kotulla et al., 2019).

Underpinning the identification of benefits of walking is the need to find accurate ways of measuring the walking undertaken by citizens, as a first step in underpinning policies which lead to greater activity. Merom and Korycinski (2017) summarise the six purposes of walking in a taxonomy which subdivides walking into primary and secondary purposes (Table 3.1). This shows that walking has multifaceted purposes, and that understanding the reasons for walking is important to its measurement. In terms of the science, the measurement of walking is most accurate using accelerometers, increasingly used in portable devices such as smartphones whose penetration is close to saturation and with increasing adoption of wearable devices following consumer preferences towards the technology. Research to identify how walking is measured goes beyond the use of technically based devices and includes objective self-report measures (e.g., asking for the number of kilometres walked) as well as more qualitative approaches (to establish how respondents feel about the distance). It is certainly the case that the ability to measure walking accurately underpins the accurate monitoring of the prevalence of walking and is a prerequisite for formulating policy to increase physical activity.

Table 3.1. A Taxonomy of Walking.

Type			
	Walking – Primary Purpose		
Purpose	For exercise, to get fit	Recreational walks, for pleasure	
Domain	Discretionary (unrestricted) leisure time	Discretionary leisure time	
Intensity	Moderate to high	Self-selected	
	Walking – Secondary Purpose		
Purpose	For travel, to get to and from places	Utilitarian (errands, shopping, personal business)	General ambulation, pacing
Domain	Restricted (at work, household duties) and discretionary time	Restricted and discretionary time	Restricted and discretionary time
Intensity	Self-selected	Self-selected	Low

Source: Merom and Korycinski (2017).

The location of walking infrastructure – whether in metropolitan or regional areas or near job sites versus residential areas – should be a key consideration, not least given that working from home has become more popular since the pandemic and resulted in changed travel patterns among working people (Bick et al., 2023). In this sense, enhancing walking infrastructure in residential neighbourhoods may yield greater benefits than before the COVID-19 pandemic. Besides location, the timing of infrastructure development is also important. Although there are often more discussions about the provision of walking infrastructure in brownfield sites, including such infrastructure in greenfield development is also important (Kent et al., 2019). For instance, installing proper pavements at the onset of greenfield projects not only demonstrates thorough urban and transportation planning but also helps in reducing potential future costs compared to the high expenses of modifying an existing built environment.

User characteristics should also be taken into account; emphasising the construction of walking infrastructure can have a more significant effect on the promotion of the mode for people who view poor or absent infrastructure as a major obstacle of conducting walking behaviours (Pearson et al., 2022). The concept of 'social infrastructure' or 'sympathetic infrastructure'[1] suggests that infrastructure should be designed and delivered in a way that supports the long-term well-being of individuals and communities. This includes considerations for minority user groups, such as people with disabilities, and can serve as a valuable guide in the development of walking infrastructure. Additionally, the development of walking infrastructure needs to consider those new modes of transport that have emerged in recent years, such as e-bikes and e-scooters.

Besides location and timing, the design of walking infrastructure is also important. Creating pedestrian-friendly environments involves designing the built environment in ways that enhance the walking experience, thereby encouraging people to choose walking over other modes, most notably driving. This includes two main aspects: ensuring pedestrian safety and making walking an enjoyable experience. First, improved infrastructure can make walking trips safer, leading to increased usage. Traffic calming measures, such as the implementation of speed bumps and designated pedestrian zones, can effectively reduce vehicle speeds and decrease the incidence of pedestrian injuries (Retting et al., 2003). Also, enhanced street lighting and surveillance on streets, including video surveillance (active surveillance) and the presence of other people (passive surveillance), can substantially influence the perceived safety of walking, and this is especially true for women who may feel more vulnerable in public spaces (Clancey et al., 2012; Koskela, 2002; Painter & Tilley, 1999; see also Chapter 16). The concept of passive surveillance refers to the informal observation that occurs when people are simply present in a space; the awareness that one is being watched by others tends to deter criminal activity and enhances a sense of security among pedestrians. Consequently, environments that are bustling with people are generally perceived as safer. It is also worth mentioning how terrorism reshapes urban atmospheres by imposing new security measures that change the way pedestrians move through and interact with the walking space (Fregonese et al., 2024). These measures, such as surveillance, physical barriers, and visible police presence, create a heightened sense of threat that lingers even when there is no immediate danger. Pedestrians may feel more anxious or self-conscious in this environment.

Second, walking experiences should be enjoyable. Enablers of more comfortable walking environments include more parks, shaded areas, and other types of green spaces (Frumkin, 2003). Potential barriers might involve negative environmental exposure such as traffic-related air pollution (Leung & Le, 2019), noise (SPIN, 2021), bad weather (Rérat, 2019), and injury risks (Sikka et al., 2019). Importantly, the motivation to reduce the greenhouse gas emissions may also encourage people to opt for walking (Gebhardt et al., 2022). And last but not least, a well-designed environment is helpful in building a welcoming social environment for pedestrians, free from hostility from car or bicycle users. Properly planned spaces can prevent car users from viewing pedestrians as obstacles, thereby promoting harmony among different transportation modes. In the urban planning realm, there has long been a concept of designing cities around walkability (Southworth, 2005). Nowadays, urban planners and professionals advocate walkability to encourage a slower pace of life and to counteract the dominance of higher-speed transport and the pursuit of efficiency. This advocacy is not about reverting to the walkable cities of the 19th century, which were shaped by limited transport options but rather about finding ways to integrate walkability into contemporary transport systems. One example of such an initiative is the '20-minute neighbourhood' concept implemented in Portland, USA (The Portland Plan, 2012): the rationale for the 20-minute benchmark is that it represents the maximum time people are generally willing to walk for everyday needs, such as grocery shopping. The case study in Box 3.1 explores the related '15-Minute City' concept in Paris, showcasing it as an exemplary model for designing walkable cities.

Box 3.1. 15-Minute City in Paris

The '15-minute city' concept was introduced by Carlos Moreno (Moreno et al., 2021) in 2016. Following the 2015 Paris Climate Change Conference emphasising the low-carbon city, the idea is to create neighbourhoods where all essential residents' needs are within reach of 15 minutes on foot or by bike (Khavarian-Garmsir et al., 2023).

The '15-minute city' initiative, launched by Paris Mayor Anne Hidalgo during her 2020 re-election campaign, aims to foster diverse and self-sufficient communities. In this plan, essential services such as housing, employment, food, recreation, and other amenities are designed to be accessible within a 15-minute walk or bike ride, eliminating the dependence on cars. This concept is founded on four key principles: density, proximity, diversity, and digitalisation (Moreno et al., 2021), which speak to the classic '5D' framework of travel demand.

The programme was initiated before the COVID-19 pandemic but gained significant traction as a model for urban recovery in the post-pandemic world (C40 Cities, 2021). The enforced lockdowns led residents to experience city life more intimately at a local level, fuelling a desire for this accessible urban design. The '15-minute city' concept has since transcended Paris, inspiring a global movement in many cities to adopt similar initiatives, such as Shanghai, Ottawa, and Melbourne (Gongadze & Maassen, 2023). However, the concept has also faced some resistance (Horton, 2024), indicating mixed reception levels and the need to cater to local backgrounds.

Decisions on infrastructure must be based on reliable, high-quality data. Planning and policy decisions commonly need to be based on understanding data, while current data collection of walking is often incomplete or inconsistent (Gerike et al., 2019). Enhancements in data collection could be achieved by sourcing data from social media, insurance companies, police reports, and other organisations, as well as by standardising data collection practices to ensure the quality and usefulness of the data. For example, in Australia, the most reliable and broadly accessible quantitative indicator of walking activity is the census' journey-to-work data, but this is limited so academic studies also largely rely on data derived from surveys and interviews. As mentioned earlier, accelerometers, now commonly integrated into smartphones and wearable devices, offer the current most precise method for measuring walking (although there still can be some issues with accuracy – see Merom & Korycinski, 2017). Policymakers have significant potential to leverage these data to monitor walking behaviours. However, relying solely on these technologies risks excluding those without access, highlighting the importance of addressing the digital divide. This emphasises the importance of self-reported measures and qualitative data, capturing the social and behavioural contexts of walking that purely technical measures might miss. Considering the significant role that walking plays in the transport system (Zhou et al., 2022), it is important to acquire more effective data on the sector.

Cultivating a Culture of Walking

From a subjective perspective, why do people choose to walk? To begin with, many people walk to improve or maintain their health. Walking is a low-impact activity that is accessible to a wide range of fitness levels and ages. Second, walking is often chosen because it is one of most accessible and affordable forms of transportation. Those without access to private vehicles or who live in areas with limited public transport options might rely heavily on walking to meet daily needs, such as getting to work, accessing services, or running errands. Third, some individuals walk as part of a conscious effort to reduce their environmental footprint, while others treat walking as a social activity to meet friends, engage in community life, and enjoy leisure time. In many cultures, walking is deeply tied to daily social practices, with individuals meeting, talking, and engaging with their environment through walking (Harries & Rettie, 2016). Lastly, in densely populated urban areas, walking is often the most practical way to get around.

Short distances between destinations, traffic congestion, and parking challenges make walking the quickest and most efficient option for many residents. Importantly, besides being only a mode of transport, Middleton (2009) positions walking as a practice through which people navigate and perceive time and space in different ways. She highlights that while transport policies often emphasise walking as timesaving and efficient, the actual experience of walking involves more complex temporalities, such as the 'event' of waiting, and emotional and sensory engagements with the environment. This calls for policymaking to recognise walking as a practice that supports everyday life coordination, emotional well-being, and community interaction, rather than simply focusing on its time-saving or health benefits.

Conversely, why do people choose not to walk? Where it is because they think it is not safe enough to do so (i.e., a perceived barrier), as well as increasing the safety of the external environment, encouraging behaviour change in favour of walking can also be influenced by 'messaging' (Mulley & Ho, 2017; Victorian Health Promotion Foundation, 2021). Here 'messaging' means advocacy to the public, via information-pass-on methods such as marketing campaigns, workshops and advertising. A behaviour change model developed by the University of Westminster (Darnton, 2008) acknowledges that persuasion is a key component in instigating behaviour change. For example, having a social model (such as a movie star) persuading people of the benefits of walking 10 minutes a day might do better than an academic publication on walking benefits (which may not be seen by many people). Moreover, health apps on devices such as the Apple Watch that share walking step data with friends or offer virtual or tangible rewards for achieving fitness goals have been shown to encourage physical activity (Yerrakalva et al., 2019). These tools harness the power of social influence and personal motivation to promote healthier behaviours.

Yet walking is valued differently across various demographics. As such, it is important to consider the user population group, in terms of gender, age, income levels, physical conditions (including those reliant on paratransit – see Chapter 8), as well as personal choice. There is a large and growing body of literature focusing on how vulnerable groups view and make more use of walking, such as children (Sener et al., 2019), women, and the elderly (Curl & Mason, 2019; Marquet et al., 2017; Shaer & Haghshenas, 2021). For the elderly, walking is not only a 'healthy choice' but sometimes a must-have tool to move around when people relinquish their driving licence at a certain age. School-aged teenagers and university students have also increasingly been the focus of recent studies (Borner et al., 2018; Campisi et al., 2022). Biehl et al. (2019) touch on the idea of a local mobility culture to uncover the way that active travel can fit with diverse communities, although they warn that not every community is ready to embrace the general trend of shifting away from private cars to more active transport modes.

In the real estate market, housing within walking distance of public transport, or within a higher density community, is often less affordable. This means low-income groups often need to travel a longer distance to destinations, maybe with less possibility of walking. From another perspective, providing high-quality public transport can support a broader effort to promote walking, as people may

The evidence for action in this context is growing with different jurisdictions taking various steps in this direction. Globally, walking strategies that exist highlight walking as an essential mechanism for reducing greenhouse gas emissions. Translating these into actions have taken a number of forms. In Scotland, for example, the *National Walking Strategy Action Plan* (Scottish Government, 2020) aims to make walking and cycling the natural choice for short journeys. The plan includes measures such as developing a daily walking culture, creating walkable spaces and places, supporting health and well-being, enhancing community cohesion, boosting the economy, improving education and productivity, and green infrastructure and path networks. It details specific actions and collaborative efforts across multiple sectors and governmental levels, demonstrating a holistic approach aimed at transforming Scotland into a healthier, more walkable, and environmentally sustainable nation.

In Australia, the City of Sydney's (2022) *Walking Strategy and Action Plan* recognises that encouraging a shift from private vehicle use to walking could significantly reduce air pollution in local areas, enhance air quality and increase health level for Sydneysiders. The measures identified included enhancing urban infrastructure, improving public spaces and accessibility, building a strong walking culture, safety and comfort measures, and connecting key areas. More proactively, San Francisco's WalkFirst initiative in the United States is a data-driven process to prioritise five years of capital improvements aimed at achieving walking objectives and fostering a safer environment for pedestrians (San Francisco Municipal Transportation Agency, 2010). Centred on enhancing pedestrian safety through a blend of measures including engineering, enforcement and education, the plan aims to design streets as valuable public spaces, enhancing the pedestrian environment by improving pavements, pedestrian crossings, and the overall streetscape. Many other examples exist, and perhaps one of the best known is the High Line in New York City. Transformed from an abandoned elevated railway, it demonstrates how cities can repurpose existing infrastructure to create a walkable environment in dense urban spaces (High Line, 2024; Pentagram, 2020; Riggs & Steiner, 2017).

There are also cities that have consistently presented themselves as leading examples of effective walking interventions. Freiburg, in southwest Germany, has long been one of Europe's most pedestrian-friendly cities (Riggs & Steiner, 2017). The city established its first pedestrian zone comprising five streets in 1949, contributing to its recovery from World War II (Lennard & Lennard, 1995). Over the years, the expansion of car-free zones has reinforced Freiburg's reputation as a compact, walkable holiday destination with low ownership of cars (Kelemen, 2014).

All of this said, a challenge for quantifying the contribution of walking to the net-zero agenda lies in the difficulty in data collection and the challenges in measuring the environmental benefits of walking. Advancements in evaluating walkability and its impact do hold promise for addressing this issue in the future – for example, a novel walkability index specifically designed for World Heritage Cities in developing countries, focusing on connectivity, proximity, land-use mix, and retail density (Valverde-Caballero et al., 2024) – but in the final analysis, walking is inescapably a pathway to provide a significant contribution to the achievement of net-zero targets. Providing there is a coordinated effort to overcome existing barriers and to implement effective strategies, increased walking can help with environmental sustainability, social equity, health, and economic activity. As cities continue to grow and seek sustainable solutions, walking must be central to the discourse on urban mobility, demanding a significant shift in policy but also a transformation in public culture and mindset.

NOTES

1. 'Social infrastructure is comprised of the facilities, spaces, services and networks that support the quality of life and well-being of our communities. It helps us to be happy, safe and healthy, to learn, and to enjoy life' (Infrastructure Australia, 2019, p. 388).
2. 'Traffic violence' is a term used to describe serious injuries and death on roads caused by automobiles.

REFERENCES

34th Ave Open Streets Coalitions. (2023). *Welcome to the official 34th Ave Open Streets Coalitions website!* https://www.34aveopenstreets.com/

Agrawal, W. A., Schlossberg, M., & Irvin, K. (2008). How far, by which route and why? A spatial analysis of pedestrian preference. *Journal of Urban Design*, *13*(1), 81–98.

Aliyas, Z. (2019). Built environment correlates of walking for recreation or exercise. *Journal of Public Health*, *27*(3), 349–356.

Bartolini, B. (2022). *Story of a street*. https://urbanomnibus.net/2022/07/story-of-a-street/

Bauman, A., Bull, F., Chey, T., Craig, C. L., Ainsworth, B. E., Sallis, J. F., Bowles, H. R., Hagstromer, M., Sjostrom, M., Pratt, M., & The, I. P. S. G. (2009). The international prevalence study on physical activity: Results from 20 countries. *International Journal of Behavioral Nutrition and Physical Activity*, *6*(1), 21.

Beck, M. J., & Hensher, D. A. (2020). Insights into the impact of COVID-19 on household travel and activities in Australia – The early days under restrictions. *Transport Policy*, *96*, 76–93.

Bedard, C., King-Dowling, S., Obeid, J., Timmons, B. W., & Ferro, M. A. (2022). Correlates of moderate-to-vigorous physical activity in children with physical illness and physical–mental multimorbidity. *Health Education & Behavior*, *49*(5), 780–788.

Bick, A., Blandin, A., & Mertens, K. (2023). Work from home before and after the COVID-19 outbreak. *American Economic Journal: Macroeconomics*, *15*(4), 1–39.

Biehl, A., Chen, Y., Sanabria-Véaz, K., Uttal, D., & Stathopoulos, A. (2019). Where does active travel fit within local community narratives of mobility space and place? *Transportation Research Part A: Policy and Practice*, *123*, 269–287.

Borner, K. B., Mitchell, T. B., Carlson, J. A., Kerr, J., Saelens, B. E., Schipperijn, J., Frank, L. D., Conway, T. L., Glanz, K., Chapman, J. E., Cain, K. L., & Sallis, J. F. (2018). Latent profile analysis of young adolescents' physical activity across locations on schooldays. *Journal of Transport & Health*, *10*, 304–314.

Buehler, R., & Pucher, J. (2021). COVID-19 impacts on cycling, 2019–2020. *Transport Reviews*, *41*(4), 393–400.

C40 Cities. (2021). *C40 Mayors' statement for a green and just recovery*. https://www.c40.org/wp-content/uploads/2021/07/2684_C40_GREEN_RECOVERY_STATEMENT.original.pdf

Campisi, T., Nahiduzzaman, K. M., Nikiforiadis, A., Stamatiadis, N., & Basbas, S. (2022). Will the effects of COVID-19 on commuting and daily activities of the university students be maintained? Evidence from a small town in Sicily. *Sustainability*, *14*(10), 5780. https://www.mdpi.com/2071-1050/14/10/5780

Carmona, M. (2019). Place value: Place quality and its impact on health, social, economic and environmental outcomes. *Journal of Urban Design*, *24*(1), 1–48.

Cervero, R., & Kockelman, K. (1997). Travel demand and the 3Ds: Density, diversity, and design. *Transportation Research Part D: Transport and Environment*, *2*(3), 199–219.

City of Sydney. (2022). *Walking strategy and action plan*. https://www.cityofsydney.nsw.gov.au/-/media/corporate/files/2020-07-migrated/files_w/walking-strategy-final-for-web.pdf

Clancey, G., Lee, M., & Fisher, D. (2012). Crime prevention through environmental design (CPTED) and the New South Wales crime risk assessment guidelines: A critical review. *Crime Prevention and Community Safety*, *14*, 1–15.

Conn, V. S., Chan, K., Banks, J., Ruppar, T. M., & Scharff, J. (2014). Cultural relevance of physical activity intervention research with underrepresented populations. *International Quarterly of Community Health Education*, *34*(4), 391–414.

Curl, A., & Mason, P. (2019). Neighbourhood perceptions and older adults' wellbeing: Does walking explain the relationship in deprived urban communities? *Transportation Research Part A: Policy and Practice*, *123*, 119–129.

Cusack, M. (2021). Individual, social, and environmental factors associated with active transportation commuting during the COVID-19 pandemic. *Journal of Transport & Health*, *22*, 101089.

Davis, A., & Parkin, J. (2015). Active travel: Its fall and rise. In H. Barton, S. Thompson, S. Burgess, & M. Grant (Eds.), *The Routledge handbook of planning for health and well-being* (pp. 108–120). Routledge.

Darnton, A. (2008). *Reference report: An overview of behaviour change models and their uses*. https://www.gov.uk/government/publications/government-social-research-behaviour-change-guidance-for-social-researchers-in-government

Davis, A., & Parkin, J. (2015). Active travel: Its fall and rise. In *The Routledge handbook of planning for health and well-being*. Routledge.

Department for Transport. (2022). *TAG unit A5.1 active mode appraisal*. https://assets.publishing.service.gov.uk/media/63a32b698fa8f53911cfcd36/TAG_Unit_A5.1_-_Active_Mode_Appraisal_Nov_2022_Accessible_v1.0.pdf.pdf

Ewing, R., & Cervero, R. (2001). Travel and the built environment: A synthesis. *Transportation Research Record*, *1780*(1), 87–114.

Ewing, R., & Cervero, R. (2010). Travel and the built environment: A meta-analysis. *Journal of the American Planning Association*, *76*(3), 265–294.

Fregonese, S., Simpson, P., Masson, D., Runkel, S., Benjamin, C. A., Berlin, S., Ciax, K., & Drongiti, A. (2024). Atmospheric geographies of (counter)terrorism. *Progress in Human Geography*, *48*(6), 752–778.

Frumkin, H. (2003). Healthy places: Exploring the evidence. *American Journal of Public Health*, *93*(9), 1451–1456.

Gebhardt, L., Ehrenberger, S., Wolf, C., & Cyganski, R. (2022). Can shared E-scooters reduce CO_2 emissions by substituting car trips in Germany? *Transportation Research Part D-Transport and Environment*, *109*, 17.

Gerike, R., de Nazelle, A., Wittwer, R., & Parkin, J. (2019). Special Issue: "Walking and cycling for better transport, health and the environment". *Transportation Research Part A: Policy and Practice*, *123*, 1–6.

Gongadze, S., & Maassen, A. (2023). *Paris' vision for a '15-minute city' sparks a global movement*. https://www.wri.org/insights/paris-15-minute-city

Harries, T., & Rettie, R. (2016). Walking as a social practice: Dispersed walking and the organisation of everyday practices. *Sociology of Health & Illness*, *38*(6), 874–883.

High Line. (2024). *High Line*. https://www.thehighline.org/

Horton, H. (2024). *Why has the '15-minute city' taken off in Paris but become a controversial idea in the UK?* https://www.theguardian.com/cities/2024/apr/06/why-has-15-minute-city-taken-off-paris-toxic-idea-uk-carlos-moreno

Infrastructure Australia. (2019). *Australian infrastructure audit 2019 – Social infrastructure*. https://www.infrastructureaustralia.gov.au/australian-infrastructure-audit-2019-social

Kelemen, R. D. (2014). *Lessons from Europe? What Americans can learn from European public policies*. CQ Press.

Kent, J. L., Mulley, C., & Stevens, N. (2019). Transport and wellbeing in a newly constructed greenfield estate: A quantitative exploration of the commuting experience. *Journal of Transport & Health, 13*, 210–223.

Khavarian-Garmsir, A. R., Sharifi, A., & Sadeghi, A. (2023). The 15-minute city: Urban planning and design efforts toward creating sustainable neighborhoods. *Cities, 132*, 104101.

Kim, H., Seok, H., Iris You, S., & Lee, C. (2020). An empirical analysis for mode choice in a short-distance trip with personal rapid transit. *Journal of Advanced Transportation, 2020*, 1–13.

Koschinsky, J., Talen, E., Alfonzo, M., & Lee, S. (2017). How walkable is Walker's paradise? *Environment and Planning B: Urban Analytics and City Science, 44*(2), 343–363.

Koskela, H. (2002). Video surveillance, gender, and the safety of public urban space: "Peeping Tom" goes high tech? *Urban Geography, 23*(3), 257–278. https://doi.org/10.2747/0272-3638.23.3.257

Kotulla, T., Denstadli, J. M., Oust, A., & Beusker, E. (2019). What does it take to make the compact city liveable for wider groups? Identifying key neighbourhood and dwelling features. *Sustainability, 11*(12), 3480. https://www.mdpi.com/2071-1050/11/12/3480

Kowaleski-Jones, L., Zick, C., Smith, K. R., Brown, B., Hanson, H., & Fan, J. (2018). Walkable neighborhoods and obesity: Evaluating effects with a propensity score approach. *SSM-Population Health, 6*, 9–15.

Lennard, S. H. C., & Lennard, H. L. (1995). *Livable cities observed: A source book of images and ideas for city officials, community leaders, architects, planners and all other committed to making their cities livable*. Gondolier Press.

Leung, A., & Le, T. P. L. (2019). Factors associated with adolescent active travel: A perceptive and mobility culture approach – Insights from Ho Chi Minh City, Vietnam. *Transportation Research Part A: Policy and Practice, 123*, 54–67.

Litman, T. (2017). Economic value of walking. In C. Mulley, K. Gebel, & D. Ding (Eds.), *Walking: Connecting sustainable transport with health* (pp. 81–98). Emerald Publishing Limited.

Loukaitou-Sideris, A. (2020). Special issue on walking. *Transport Reviews, 40*(2), 131–134.

Marquet, O., Hipp, J. A., & Miralles-Guasch, C. (2017). Neighborhood walkability and active ageing: A difference in differences assessment of active transportation over ten years. *Journal of Transport & Health, 7*, 190–201.

Marsden, G., & Docherty, I. (2021). Mega-disruptions and policy change: Lessons from the mobility sector in response to the Covid-19 pandemic in the UK. *Transport Policy, 110*, 86–97.

Merom, D., & Korycinski, R. (2017). Measurement of walking. In C. Mulley, K. Gebel, & D. Ding (Eds.), *Walking: Connecting sustainable transport with health* (pp. 11–39). Emerald Publishing Limited.

Middleton, J. (2009). 'Stepping in Time': Walking, Time, and Space in the City. *Environment and Planning A: Economy and Space, 41*(8), 1943–1961.

Moreno, C., Allam, Z., Chabaud, D., Gall, C., & Pratlong, F. (2021). Introducing the "15-minute city": Sustainability, resilience and place identity in future post-pandemic cities. *Smart Cities, 4*(1), 93–111. https://www.mdpi.com/2624-6511/4/1/6

Mulley, C., & Ho, C. (2017). Understanding the determinants of walking as the basis for social marketing public health messaging. In C. Mulley, K. Gebel, & D. Ding (Eds.), *Walking: Connecting sustainable transport with health* (pp. 41–59). Emerald Publishing Limited.

Neunuebel, C. (2023). *What the World Bank's country climate and development reports tell us about the debt-climate nexus in low-income countries*. https://www.wri.org/technical-perspectives/what-world-banks-country-climate-and-development-reports-tell-us-about-debt

New York City Department of Transportation. (2021). *The future of the 34th Avenue Open Street*. https://www.nyc.gov/html/dot/downloads/pdf/34th-ave-open-street-cb3-full-board-oct2021.pdf

New York City Department of Transportation. (2023). *Open streets*. https://www.nyc.gov/html/dot/html/pedestrians/openstreets.shtml

Nieman, D. C., & Wentz, L. M. (2019). The compelling link between physical activity and the body's defense system. *Journal of Sport and Health Science, 8*(3), 201–217.

Painter, K., & Tilley, N. (1999). *Surveillance of public space: CCTV, street lighting and crime prevention* (Vol. 10). Criminal Justice Press Monsey.

CHAPTER 4

THE ROLE OF HUMAN-SCALE MOBILITY IN DECARBONISING TRAVEL

John Parkin

University of the West of England, UK

ABSTRACT

This chapter identifies the 'carbon turn' in policy needed to create networks of routes that are comfortable and attractive for carbon-efficient human-scale mobility. First, the nomenclature and development of the full breadth of electric and non-electric human-scale mobility is reviewed, and the nature of human-scale vehicles is defined as including cycles and other vehicles of similar scale, such as e-scooters. Lifecycle carbon emissions in grams of carbon dioxide equivalent per passenger kilometre are summarised for human-scale mobility and compared with other private and public means of travel. Cycling emits around a tenth of the carbon dioxide equivalent emissions per passenger kilometre of car travel. Hire cycles and e-scooters emit 30%–85% of the amount emitted by cars and taxis. Estimating lifecycle carbon emissions is complex, and methods of estimation will continue to be refined. More important than refining appraisal methods is the need for a carbon turn in policy to take place so that transport investment focuses on reducing carbon emissions. Using Wales as the example, such a turn is demonstrated as being possible through firm legislative foundations and internally consistent strategy that results in specific policy, which then leads to investment directed at creating a human-scale infrastructure. The infrastructure required for good quality human-scale mobility environments includes comprehensive networks of routes that are comfortable and attractive and that separate human-scale vehicles from both pedestrians

Towards Transport Net Zero
Transport and Sustainability, Volume 20, 45–62

Published under exclusive licence by Emerald Publishing Limited
ISSN: 2044-9941/doi:10.1108/S2044-994120250000020004

and motor traffic. Current design guidance needs to be reviewed to ensure it is suitable for the widest range of wheeled human-scale mobility.

Keywords: Human-scale mobility; cycling; e-scooters; decarbonisation; transport policy and strategy; human-scale travel infrastructure

INTRODUCTION

Defining Human-scale Mobility

Cycling uses a cranked, wheeled vehicle. Walking uses the feet. For both means of travel the power comes from human muscle. Walking and cycling are neutral descriptions in the sense they are gerunds based on the verb that describes the action. Cycling and walking are often grouped as though they were a single means of travel, and they are described using a range of adjectives. A Cycling and Society Research Group email group discussion in July 2023 noted the use of the following adjectives used before the word 'travel' to describe cycling and walking: active, gentle, green, light, slow, sustainable, and sweet (akin to the Italian phrase *la dolce vita*). These adjectives describe specific characteristics of travel, rather than the means of travel. Hence, none of them are likely to be ubiquitously suitable for use in policy or planning.

Cook et al. (2022) note the limitations of one of those terms, active travel, to describe cycling and walking. They note the term's implied boundaries and exclusions; for example, manual wheelchair users, skateboarders, kick-scooterists, rollerbladers, and runners. Despite problems of definition, the Welsh Government (2013) has enshrined a definition in law for an 'active travel route' as being a route 'which may lawfully be used by, or by any description of, walkers and cyclists'. They leave open to interpretation the means of travel that may be similarly described.

For devices and vehicles that weigh up to 350 kg and are human powered, sometimes with electric-assist, the International Transport Forum (ITF, 2020a) uses the umbrella term 'micromobility'. This has an unfortunate connotation of 'thousandth-scale' (i.e., 10 to the minus three, 10^{-3}, in mathematical terms), and hence an implicit comparison with an unnamed 'normal' scale mode. They identify micromobility devices and vehicles as being: bicycles, cargo cycles and e-cargo cycles, pedal-assisted e-cycles (slow-pedelecs with power cut-off at 25 km/h, and speed-pedelecs with power cut-off at 45 km/h), throttle-controlled e-cycles, mobility scooters, standing (kick or push) scooters, e-scooters, skateboards, electric skateboards, self-balancing boards (e.g., hoverboards), one-wheel self-balancing boards, electric unicycles, roller skates and electric skates, and mopeds. The legal status of the vehicles differs by country; for example, in the United Kingdom, privately owned e-scooters are illegal: their only legal use is when hired from one of the e-scooter trial schemes that have been taking place since 2021.

The ITF groups micromobility vehicles in four ways: first, a binary division based on mass (under 35 kg or 35–350 kg), and second a binary division based on the level of power-assist (human-powered optionally with power-assist up to 25 km/h or powered with speeds between 25 and 45 km/h). Human-scale vehicles may be privately owned or available for long-term or, more typically, short-term hire. Short-term hire is usually via an application on a mobile telephone, and the price paid may include a flat rate charge plus a time charge. There are share schemes in many European, North American, and Southeast Asian cities for bicycles, e-cycles, e-scooters, and, increasingly, e-cargo cycles. Flower (2022) extends the ITF categorisation to consider range (for example, cycle journeys are often longer than e-scooter journeys) and provides a more refined sub-division for speed that relates to UK regulations for electric wheelchairs and mobility scooters as follows: Class 1 (lightweight up to 4 mph/6.4 km/h), Class 2 (up to 4 mph/6.4 km/h), Class 3 (4 mph on footways which may be switched to up to 8 mph/12.8 km/h for use on carriageways).

Clayton et al. (2017) further extend classifications for cycles and adapted cycles beyond the usual two-wheeled pedal-cranked vehicle. They identify the following and combinations thereof: solo and tandem cycles, tricycles, hand-cycles, recumbent cycles, mixed recumbent and upright cycles, wheelchair clip-on cycles, and electric-assisted cycles. Extending in a different direction, Cox and Van De Walle (2016) define velo-mobiles (often called human-powered vehicles or HPVs) as multi-wheeled cycles which have enclosing aerodynamic bodywork. Based on mechanical and design characteristics, they trace a non-linear and branching pathway of the development of crank-propelled vehicles that have come to be seen as either cycles or velo-mobiles. When considering everyday travel, velo-mobiles should be, but are often not, included in vehicle classifications. In yet a different direction, Jones and Chatterjee (2023) edited a special issue of the *Active Travel Studies* journal on e-micromobilities. They therefore implicitly equate e-micromobilities with active travel, and their editorial provided arguments to support their case. They are honest enough to note that their stance received some criticism from paper reviewers.

Arguably then, there is an unresolved matter concerning a suitable group noun for the variety of means of travel described above. This is because there is no term used to date which commands complete acceptance. It is logical to reject terms that have limited purview (e.g., active travel) or describe a characteristic rather than the means of travel (e.g., sustainable travel). It is also logical to reject the term ‘micromobility’ because it defines vehicle size using an implausibly large ratio in relation to an unstated norm. Finally, to be inclusive, it is appropriate to avoid separately defining some vehicles as being adapted.

Human-centred planning comes to us as a philosophy and a term from the renowned urbanist and architect Jan Gehl (Musselwhite, 2023). In relation to travel research, the term ‘human-scale’ already has pedigree (e.g., Appleyard, 2016; Mladenović et al., 2021). Human-scale is a group noun that is inclusive, right-scale, and all-encompassing. Some may suggest human-scale is defining a characteristic of the travel, and that therefore places it at the same disadvantage as other descriptions noted above. However, as a group noun, its nature is inevitably to be descriptive of the group, and this is akin, for example, to individual modes such as bus and rail having a description which describes their nature: viz. public

Table 4.1. Carbon Dioxide Equivalent Emissions of Selected Urban Modes of Travel.

Vehicle	Lifespan (Vehicle Kilometres)	Average Passenger Load	Carbon Emissions Component (gCO_2 equivalent per Passenger Kilometre)				
			Vehicle	Fuel	Operations	Infrastructure	Total
Bicycle	13,440	1	7	–	–	9	16
E-cycle	13,440	1	13	12	–	9	34
Bicycle shared	5,510	1	23	–	25	9	57
E-cycle shared	5,510	1	37	12	25	10	84
E-scooter	6,600	1	26	6	–	9	41
E-scooter shared	5,715	1	66	7	25	9	107
Car – ICE	181,500	1.5	24	126	0	12	162
Car – BEV	181,500	1.5	42	72	0	12	126
Taxi – BEV	339,360	0.73	50	90	83	26	249
Pre-booked PHV – ICE	339,360	0.95	22	122	112	20	276
Pre-booked PHV – BEV	339,360	0.95	39	69	64	20	192
Bus – ICE	396,000	17	8	72	8	4	92
Bus – BEV	396,000	17	14	46	5	4	69
Metro/urban train	2,640,000	190	2	53	0	11	66

Source: Based on data from International Transport Forum (2020b).
Note: ICE is internal combustion engine, BEV is battery electric vehicle, PHV is private hire vehicle, sometimes called ride-sourcing.

The upper part of the table shows six human-scale modes, both privately owned and shared. These may be compared with private and public transport modes in the lower part of the table. While these internationally derived values are useful for a high-level comparison, local values for countries and/or cities need to be estimated for appraisal of individual schemes in specific localities. This is because of variability in emissions from, for example, the generation and distribution in the energy supply chain and, for share schemes, different approaches to operational servicing.

A privately owned bicycle has the lowest emission level (16 gCO_2eq/pass.km) and is 6%–13% of the emissions of cars, taxis, and PHVs. No fuel component is provided for cycling in the ITF data. Metabolic equivalent of task (MET) is equal to 1.162 W/kg of body mass and one MET is equivalent to sitting and resting. Walsh et al. (2008) estimate that cycling at up to less than 16 km/h has an MET of four resulting in cyclists breathing out 5 gCO_2eq/pass.km. Haskell et al. (2007) suggest an MET of six for cycling at up to 19 km/h (and an MET of 3.3 for walking at 4.8 km/h).

Walsh et al. (2008) do not consider the carbon emissions of food production. Assuming a human body efficiency of 25%, energy expenditure at 4 MET for a kilometre at 16 km/h requires 75 kcal. Scarborough et al. (2014) estimate carbon dioxide emissions for a standard daily diet of 2,000 kcal eaten in the United Kingdom at 2.89 $kgCO_2e$/day for vegans to 7.19 $kgCO_2e$/day for high-quantity meat-eaters. Simple factoring suggests food-related emissions for cycling a kilometre are in the range 108–269 gCO_2eq/pass.km. However, it is not necessarily the case that additional food is consumed to undertake daily short cycling journeys. It may be that food consumed anyway is being put to better use travelling than it might otherwise have been. Also, if the carbon emissions from food for a cyclist are to be considered, then similarly, the carbon emissions from every human in the travel industry supply chain needs to be considered: car factory workers, workers in power stations and oil refineries, taxi drivers, and so on. It is also worth bearing in mind that a direct and natural consequence of additional food consumption relating to cycling (and walking) is increased physical activity which can have large health co-benefits (Woodcock et al., 2009).

ITF estimates the infrastructure component for cycling at 9 gCO_2eq/pass.km. Their infrastructure estimates for all modes include embodied carbon only in materials (excluding earthworks for example), maintenance, and an assumed lifespan of 30 years for roads and 50 years for rail track. Assumptions were made about volume of vehicle traffic. Infrastructure for share schemes, such as docking stations were not estimated.

Human-scale mobility share schemes have emissions around 2.5–3.5 times greater than privately owned human-scale vehicles. The additional emissions are because of higher vehicle embodied carbon, marginal additional infrastructure requirements, and servicing. At 84 and 107 gCO_2eq/pass.km, shared e-cycles and shared e-scooters emit 30%–85% of the carbon dioxide equivalent emitted by cars, taxis, and PHVs. Shared human-scale hire schemes emit at levels on a par with public transport (ranging from around 90% to 160% of emissions of buses and urban rail per passenger kilometre). Some ideas of the possible variation in the estimates in Table 4.1 can be gleaned from inspecting other sources that have estimated emissions for specific circumstances. For cycles, but excluding the infrastructure component, Weiss et al. (2015) estimated 4–6 gCO_2eq/pass.km for cycles and 10–70 gCO_2eq/pass.km for e-cycles. For seven cities in Europe, Brand et al. (2021) estimated 5.1–5.3 gCO_2eq/pass.km for cycles and e-cycles combined (with an assumed e-cycle share of 4.5%). Again, this excludes infrastructure and assumes a lifespan of 20,000 km, i.e., about 50% longer than the ITF assumed lifespan.

Estimating for all emission components including infrastructure, de Bortoli (2021) estimates the following emissions for Paris (all in gCO_2eq/pass.km): cycles, 12; shared cycles, 33; e-sooters, 42–60; shared e-scooters, 61. In the North American context and excluding the infrastructure component, Hollingsworth et al. (2019) estimate shared e-scooter emissions of 88–126 gCO_2eq/pass.km, depending on lifespan assumptions. In Berlin, the range estimated by Severengiz et al. (2020) for shared e-scooters is 64–237 gCO_2eq/pass.km. Ishaq et al. (2022) report 21 gCO_2eq/pass.km for private e-scooters in Turin. These data show the considerable variability in estimates per kilometre travelled in carbon dioxide emission per passenger: for cycle around 16; for e-cycles around 34; for cycle and e-cycle share schemes an additional 25; for e-scooters 21–60; for shared e-scooter schemes from 61 to 237 (all in gCO_2eq/pass.km). By comparison, the UK Department for Business Energy and Industrial Strategy (BEIS) (2023) indicates fuel consumption-related emissions for an average internal combustion engine car of 166 gCO_2eq/pass.km. Despite not including fuel supply emissions, this estimate is higher than the ITF estimate which does include fuel supply. The BEIS estimate for a battery electric vehicle is 55 gCO_2eq/pass.km and does include the electricity production and transmission and distribution emissions (Chapter 5). This is similar to the ITF battery electric vehicle estimate of 72 gCO_2eq/pass.km.

Broad patterns of carbon emissions from the use of different vehicles are evident, and there will be diminishing value in continuing to refine emissions estimates per passenger kilometre, particularly when there is a great deal of variability in other aspects of appraisal and evaluation of travel schemes, such as estimating demand and diversion factors, which are discussed in the next section.

Carbon Evaluations of Interventions

The effectiveness and cost effectiveness of publicly funded transport schemes are usually appraised before implementation. That appraisal is informed by evidence of impacts based on the evaluation of schemes already in place (HM Treasury, 2020). This section discusses human-scale mobility evaluation findings that may be used to inform appraisal. A central requirement of appraisal is the estimation of demand for travel. To estimate demand, large transport schemes may be the subject of considerable modelling that estimates trip generation, distribution, mode choice, and route choice. Often, human-scale schemes will be relatively small in extent and impact. For reasons of proportionality, appraisal for such smaller schemes may adopt less detailed estimation methods for forecasting demand. A parameter often used to estimate demand is the diversion factor, which indicates the proportion of demand that provision of new infrastructure for an existing mode, or a newly introduced mode, may attract from other existing modes. These diversion factors are based on evidence from previous interventions that have been evaluated.

For cycle interventions, transport appraisal guidance published by the DfT (2023b) presents diversion factors to cycle from car of 24%, from bus and walking 14% each, with 23% of the demand being newly generated. In developing these estimates for the guidance, Clark and Parkin (2022) drew mostly on intercept surveys from seven cities with infrastructure improvements funded by the Cycle City Ambition programme that demonstrated 12%–40% of new cycle trips came from diversion from car (Sloman et al., 2021). Woodcock et al. (2014) report that 2% of new cycle trips using the cycle hire scheme in London diverted from car, reflecting the low level of car use in London. In contrast, Murphy and Usher (2015) found a diversion factor to cycle from car for the Dublin Cycle Hire scheme of 20%. Bourne et al. (2020) reviewed the literature on diversion to e-cycles and reported wide ranges: from conventional cycle, 23%–27%; from car 20%–86% and from public transport 2%–45%.

The causes of these wide-ranging estimates come from multiple sources including the following: trip length, trip purpose, trip frequency, age and gender, group (e.g., escort) versus individual trips, public transport cost and level of service, car ownership. In addition, the pre-existing cycle mode share may influence diversion to cycling resulting from an intervention such as new infrastructure. The rate of diversion will vary based on the quality of the infrastructure and its nature and setting within the network. A route, even of moderate length and good quality, within an area that is otherwise ill-served for human-scale mobility may have less effect on demand than additional links in an otherwise well-developed network.

There may be other unexpected effects of a switch in mode. For example, de Bortoli and Christoforou (2020) found that the total distance travelled by e-scooter users in Paris reduced by 38.5%. They note that average trip distances vary between modes, and that this could indicate that the distance reduction is created by a change in destination choice in combination with the change in mode. An example may be that a switch from car to scooter also encourages shopping closer to home. They also note that public transport may not be routed on the shortest path for a particular journey, and because of area wide traffic management, this may be true of a car journey too. Chatterjee et al. (2023) demonstrated variability in estimates of diversion factors for an e-scooter hire trial in the West of England Combined Authority area focused on Bristol and Bath. They carried this through to an illustration of the sensitivity of carbon emission reductions resulting from the scheme. They found the level of carbon reduction was sensitive to the balance of abstraction from walking versus car travel and to a lesser extent walking versus bus travel. They note the possibility that e-scooters may replace walk trips more frequently than car trips, and this effect would require data from trip diaries as opposed to cross-sectional survey data. Also, e-scooter trips that replace walk trips may be shorter than those that replace car trips.

Lifecycle emissions data provide for the most comprehensive estimates of carbon emissions change, and these are most relevant where the changes in demand resulting from an intervention are likely to be high. Where an intervention is modest in scale, which many human-scale interventions are, change in demand may be marginal. Thinking about a switch for occasional journeys from car to cycle, there may be no change in vehicle ownership and modest impact on infrastructure. In such cases, the main source of change in emissions will relate only to the fuel component. Proportionality suggests only fuel-related emissions would be of concern. However, the aim of policy, discussed in the next section, needs to be directed at creating significant reductions in carbon emissions from transport. This implies changes are required at a scale that is more than marginal. This would require, for example, a significant change in the quantity of comfortable and attractive human-scale infrastructure that is then the enabler of significant changes in travel choices and vehicle ownership.

E-scooter trials, even the most successful UK trial in the West of England area in which 8% of the population used an e-scooter at least once a month, may still appear to be modest in scale based on mode choice changes. Relative to other vehicles, the lifespan of an e-scooter has been short, at least to date. This means that the embodied carbon in the e-scooter as a vehicle is a significant proportion of lifecycle emissions. This suggests that an assessment dealing only with direct and creating a modal shift, with the need to reduce carbon emissions being the central reason for the need to change. By contrast, the DfT (2017, para 2.9) notes that ‘Protecting our environment and health will continue to involve considerable effort and resource and may in places constrain the choices we make’. This carefully crafted sentence implies that environmental constraints are not a prime motivator for the choices that need to be made.

A strategy needs action. As a direct and natural consequence of the strategy, Welsh Government established a panel to review the roads programme to check its alignment with the new transport strategy. This resulted in 31 of the 48 road schemes in the programme that existed at the time of the review being cancelled. The panel’s wider recommendations resulted in a change in policy for road building by the Welsh Government (2023a). Roads will continue to be considered by the Welsh Government, but only in the following circumstances: to support modal shift and reduce carbon emissions, to improve safety through small-scale changes, to adapt to the impacts of climate change, and to provide access and connectivity to jobs and centres of economic activity in a way that supports modal shift. In addition, the Welsh Government (2023b) introduced a reduced default speed limit in urban areas across Wales of 20 mph from 17 September 2023. At the same time, the Welsh Government had continued to develop sustainable transport programmes of investment not least as a result of the recommendations of two independent commissions, one for south east Wales and one for north Wales (North Wales Transport Commission, 2023; South East Wales Transport Commission, 2020). The commissions have placed walking and cycling at the top of the hierarchy in their recommendations. Political transport leadership change may create further twists in the process of making a turn.

This comparison of the now divergent transport national policies of the neighbouring countries of England and Wales demonstrates four important ways that a ‘carbon turn’ in policy can be made (at the local level, particularly mayoral authorities in England are making better progress than the DfT in promoting sustainable travel). First, to make significant change the foundations need to be carefully laid. In the case of Wales, these foundations are encapsulated in law in the future generations act. Second, strategy needs to be consistent across all domains: transport, planning, and carbon. Third, strategy needs to be developed into specific polices, such as the new roads policy that defines the circumstances in which roads may be built. Fourth, investment needs to be directed at recommendations for transport interventions that are derived from the strategy and policy.

Infrastructure for Human-scale Mobility

Historically, the needs of people walking and cycling have not been prioritised. This is evident from the way that many areas are dominated by motorised traffic. Much has been changing and developing in relation to infrastructure design for human-scale mobility in the 21st century in many developed countries. The Dutch have known of the requirements for some time: networks that are complete and as direct as possible, attractive, comfortable, and safe (CROW, 2017). They are clear that cyclists and motor vehicles have different features and characteristics and need to be separated. Similarly, there are differences between walkers and cyclists, not least in terms of relative speed (an average of around 4 km/h for walking compared with up to 20 km/h for cycling). This means walking and cycling need to be separated, especially where there are high levels of use and hence many interactions. Such good and appropriate design philosophy is now embedded in Welsh and English guidance and is being used particularly well in some larger cities such as Cardiff and London and other mayoral authority areas and other urban areas in England.

Where interactions occur, there is the risk of collision and injury. Evidence from the United States (Cicchino et al., 2021) notes a higher level of injury from e-scooter collisions on the footway than on the carriageway. This points both to a need for separate infrastructure for human-scale wheeled travel, and the risks presented to pedestrians by wheeled vehicles being in the footway environment. Currans et al. (2022) found that in preference to a carriageway, e-scooter riders may select the footway in the absence of infrastructure separated from motor traffic. Siebert et al. (2021) similarly note that riders may use 'incorrect' infrastructure if appropriate infrastructure is not in place. Chatterjee et al. (2023) evidence the barriers to access and sense of risk among e-scooter users created because of a lack of infrastructure.

Given that cyclists' and e-scooter riders' sharing with either motorised traffic or pedestrians is problematic, the question is whether they are sufficiently alike to share 'the third space' of a separated cycle track not occupied by pedestrians or motor traffic. E-cycles and e-scooters are faster uphill than cycles. Downhill there is little difference between cycles and e-cycles because e-cycle power cuts out at 15.5 mph. Cycles may be faster than e-scooters if, for example, their maximum speed is limited: this was the case in the West of England trial where speed was limited to 12.5 mph. This means that e-scooters travel at a more consistent but typically slightly lower speed than cycles and e-cycles. Chatterjee et al. (2023) report that professional stakeholders think the same infrastructure can be used for cycles and e-scooters, because cycles, e-cycles, and e-scooters are of similar size and speed. Stakeholders recognised that cycle design guidance (DfT, 2020) may need to be reviewed and updated to ensure its detail is appropriate for the lesser gyroscopic stabilising effect of the e-scooter's smaller wheel diameter, its overall smaller size and (generally slightly) lower centre of gravity. The most relevant concerns were the smoothness of running surfaces and adequacy of widths for the volume of use. Design consideration also needs to extend to junctions to create safe and timely passage of cycles and e-scooters.

For human-scale mobility to be allowed to properly flourish, a network of routes therefore needs to be created that is separated from both pedestrians and motor traffic. The Netherlands provides a good example of how such separated networks can be created (Fig. 4.1).

Fig. 4.1. A Walking and Cycling Route in Delft, the Netherlands.
Source: Author.

CONCLUSION

Transport policy and strategy needs to support travel in a way that will reduce damage to the environment, particularly through reducing carbon emissions, and that enhances social and economic well-being. Cycling emits around a tenth of the carbon dioxide equivalent emissions per passenger kilometre as car travel. Hire cycles and e-scooters emit 30%–85% of the amount emitted by cars and taxis. Hence, policy that enables human-scale mobility to flourish is important for carbon dioxide emission reduction. The better human-scale modes to promote will be the ones which have the lowest carbon emissions per kilometre and also provide co-benefits such as physical activity.

Estimating lifecycle carbon emissions is complex because of the many components, which include embodied carbon in the vehicle, fuel and the energy supply chain, operations linked with hire, and the infrastructure. Despite possible future refinements to calculation methods, there will remain wide confidence intervals for estimates of the change in emissions because of transport interventions. Diversion factors used to estimate changes in mode choice because of human-scale mobility interventions averaged at a country level will create only approximate estimates of emissions changes for specific locations. Values of variables used in the estimation process will differ even between cities within a country. This is because of the effect of local travel patterns, geography, and other existing transport provision.

While methods of estimation will continue to develop for intervention appraisal, there will be a diminishing return in the value of this refinement. It is important, however, that forecast carbon emissions from transport scheme proposals are estimated on a like-for-like basis across all types of transport schemes. All components of emissions should be included, but where this is not practical, comparisons should be made using the same number of components and the same level of detail. The same scoping rules should be applied to all types of transport scheme for deciding whether detailed carbon emissions appraisal is required. It is likely that more schemes will be in scope for appraisal that enable greater use of high carbon-emitting modes, such as the car. This appears to be the inverse of the current situation where human-scale mobility schemes are closely scrutinised, while schemes that enable more car travel may not be appraised because of a high threshold of change being adopted at the scoping stage.

A 'carbon turn' in policy needs to take place in transport to help significantly reduce carbon emissions. First, firm foundations, possibly legislative, need to be laid. Second, strategy needs to be consistent across all relevant domains. Third, strategy needs to work through into specific policy, and finally, investment needs to be directed at implementing the policy. Infrastructure should be implemented that enables both human-scale mobility and at the same time manages and controls motor vehicle travel demand. This includes setting suitably low motor traffic speed limits in urban areas. Such infrastructure and infrastructure management will help create the right conditions for a shift to lesser emitting human-scale modes. More specifically, the infrastructure that needs to be built for human-scale mobility should: (a) be separated from pedestrians, (b) be separated from fast and/or high-volume motor traffic, (c) create networks that are comprehensive, and (d) create an environment that is comfortable and attractive for riders. Design guidance needs to be reviewed and updated as necessary to accommodate the requirements of all types of wheeled human-scale mobility.

REFERENCES

Appleyard, B. (2016). New methods to measure the built environment for human-scale travel research: Individual access corridor (IAC) analytics to better understand sustainable active travel choices. *Journal of Transport and Land Use*, *9*(2), 121–145.

BEIS. (2023). *Greenhouse gas reporting conversion factors*. Department for Business Energy and Industrial Strategy. https://www.gov.uk/government/publications/greenhouse-gas-reporting-conversion-factors-2023

Bourne, J. E., Cooper, A. R., Kelly, P., Kinnear, F. J., England, C., Leary, S., & Page, A. (2020). The impact of e-cycling on travel behaviour: A scoping review. *Journal of Transport & Health*, *19*, 100910.

Brand, C., Götschi, T., Dons, E., Gerike, R., Anaya-Boig, E., Avila-Palencia, I., de Nazelle, A., Gascon, M., Gaupp-Berghausen, M., Iacorossi, F., Kahlmeier, S., Panis, L. I., Racioppi, F., Rojas-Rueda, D., Standaert, A., Stigell, E., Sulikova, S., Wegener, S., & Nieuwenhuijsen, M. J. (2021). The climate change mitigation impacts of active travel: Evidence from a longitudinal panel study in seven European cities. *Global Environmental Change*, *67*, 102224.

British Standards. (2023). *PAS 2080:2023 carbon management in buildings and infrastructure*. British Standards. Retrieved November 30, 2023, from https://knowledge.bsigroup.com/products/carbon-management-in-buildings-and-infrastructure?version=standard

Tuncer, S., Laurier, E., Brown, B., & Licoppe, C. (2020). Notes on the practices and appearances of e-scooter users in public space. *Journal of Transport Geography*, *85*, 102702.

Walker, I., Tapp, A., & Davis, A. (2023). Motonormativity: How social norms hide a major public health hazard. *International Journal of Environment and Health*, *11*(1), 21–33.

Walsh, C., Jakeman, P., Moles, R., & O'Regan, B. (2008). A comparison of carbon dioxide emissions associated with motorised transport modes and cycling in Ireland. *Transportation Research Part D: Transport and Environment*, *13*(6), 392–399.

Weiss, M., Dekker, P., Moro, A., Scholz, H., & Patel, M. K. (2015). On the electrification of road transportation – A review of the environmental, economic, and social performance of electric two-wheelers. *Transportation Research. Part D, Transport and Environment*, *41*, 348–366. https://doi.org/10.1016/j.trd.2015.09.007

Welsh Government. (2013). *Active Travel (Wales) Act 2013*. https://www.legislation.gov.uk/anaw/2013/7/section/1/enacted

Welsh Government. (2015). *The well-being of future generations*. https://gov.wales/well-being-of-future-generations-wales

Welsh Government. (2021a). *Llwybr Newydd: The Wales transport strategy 2021*. https://gov.wales/llwybr-newydd-wales-transport-strategy-2021

Welsh Government. (2021b). *Future Wales: The national plan 2040*. https://gov.wales/future-wales-national-plan-2040

Welsh Government. (2021c). *Net zero Wales carbon budget 2 (2021 to 2025)*. https://gov.wales/net-zero-wales-carbon-budget-2-2021-2025

Welsh Government. (2023a). *Welsh Government response to the roads review*. https://www.gov.wales/roads-review-welsh-government-response

Welsh Government. (2023b). *Introducing default 20mph speeds*. Retrieved January 26, 2024, from https://www.gov.wales/introducing-default-20mph-speed-limits

Woodcock, J., Edwards, P., Tonne, C., Armstrong, B. G., Ashiru, O., Banister, D., Beevers, S., Chalabi, Z., Chowdhury, Z., Cohen, A., Franco, O. H., Haines, A., Hickman, R., Lindsay, G., Mittal, I., Mohan, D., Tiwari, G., Woodward, A., & Roberts, I. (2009). Public health benefits of strategies to reduce greenhouse-gas emissions: Urban land transport. *The Lancet*, *374*(9705), 1930–1943.

Woodcock, J., Tainio, M., Cheshire, J., O'Brien, O., & Goodman, A. (2014). Health effects of the London bicycle sharing system: Health impact modelling study. *BMJ*, *348*, 425.

CHAPTER 5

CHALLENGES TO ACHIEVING NET ZERO THROUGH TRANSITION TO THE ELECTRIC CAR

Graham Parkhurst

University of the West of England, UK

ABSTRACT

The transition from internal combustion engine (ICE) to electric powertrains is arguably the most important means of achieving net zero (NZ) greenhouse gas (GHG) emissions in a transport sector struggling to decarbonise. A NZ electric vehicle (EV) would be one for which all the emissions from production, servicing, and operation are either avoided or offset. This chapter considers three key challenges for a NZ transition. First, current EVs are far from carbon neutrality. Second, from a social perspective, EVs offer attractive features, but practices and norms have developed around the capabilities of the ICE car. Third, restructuring by the established auto industry to produce affordable EVs profitably is not progressing smoothly or sufficiently quickly. It is concluded that the transition may be faltering in the wealthy democracies. To recover momentum, there needs to be renewed examination of policy interventions to revitalise the transition while ensuring it is equitable. Greater alignment with wider sustainable mobility policy could also ease the transition, including encouragement for EV specifications compatible with service niches that support an integrated transport system which prioritises active and collective travel.

Keywords: Electric vehicle; sociotechnical transition; internal combustion engine vehicle; automobility; net zero; sustainable mobility

Towards Transport Net Zero
Transport and Sustainability, Volume 20, 63–82

ISSN: 2044-9941/doi:10.1108/S2044-99412025000020005

INTRODUCTION: CONTEXT OF TRANSITION TO THE ELECTRIC CAR

In 2019, the road system was responsible for over two-thirds (69% or 6 $GtCO_2eq$) of global transport emissions (Jaramillo et al., 2022; see Chapter 2). Contributions are highest in the global regions in which car dependence is greatest, particularly North America, followed by Europe and East Asia, although transport emissions from other regions will grow if current trends are not reversed. In 2022, the International Panel on Climate Change (IPCC, 2022, paragraph C.8) reaffirmed as a headline statement that 'Electric vehicles powered by low-GHG [greenhouse gas] emissions electricity have large potential to reduce land-based transport GHG emissions, on a lifecycle basis (*high confidence*)'. Indeed, EVs together with more efficient vehicles generally were identified as the demand-side contribution expected to make the greatest difference in reducing emissions from mobility.

The IPCC was responding to the status quo resulting from a 20th century in which road passenger transport became increasingly dominated by the light passenger vehicle powered by an ICE. That earlier round of technological transition first became significant in the United States from the 1910s, followed by mass adoption in Europe, and by the end of the century was well underway in nearly every industrialising nation. The nature of ICE domination has varied over time and between places. Motorisation in Europe has been characterised by the small- and medium-sized passenger car and in the United States and Australia by larger passenger cars and light trucks. In many Asian and African contexts, the motorcycle and its derived three wheelers are the key mode of transport. These vehicles were once more significant in Europe before being replaced by cars, and there are uncertainties about how far and how quickly four wheelers might, too, dominate in these motorising markets. Electric versions of two and three wheelers are beyond the scope of the current chapter, which focuses on addressing the dominating GHG emissions from 'light-duty' car and truck ICEs, but it is important to observe that 95% of current EVs are in this very light category and are making an appreciable contribution to decarbonisation (BloombergNEF, 2022).

Regarding the electric car (following common practice, referred to hereafter as the 'electric vehicle', or EV), the second half of 2024, when this chapter was drafted, was a key and dynamic moment in the global transition. By 2023, the world stock of pure battery EVs (BEVs) reached 28 million (Fig. 5.1), of which 57% were in China, around a quarter in Europe (24%), and most of the rest in the United States (12%). China's leading role reflects EVs having been a key component in strategic industrial policy since 2001, following recognition it would be extremely challenging to compete with the established global auto industry on ICE technology. From 2009, EV producers (whether Chinese or foreign subsidiaries) were a focus for public sector investment (also identified as subsidies) amounting to €25 billion by 2022. Public transport operators and private consumers benefited from lower purchase costs due to these subsidies and were also incentivised through regulatory measures such as easier vehicle registration procedures (Yang, 2023).

Norway, without a domestic automotive production industry, represents an equally striking but contrasting example, in this case of consumption-oriented

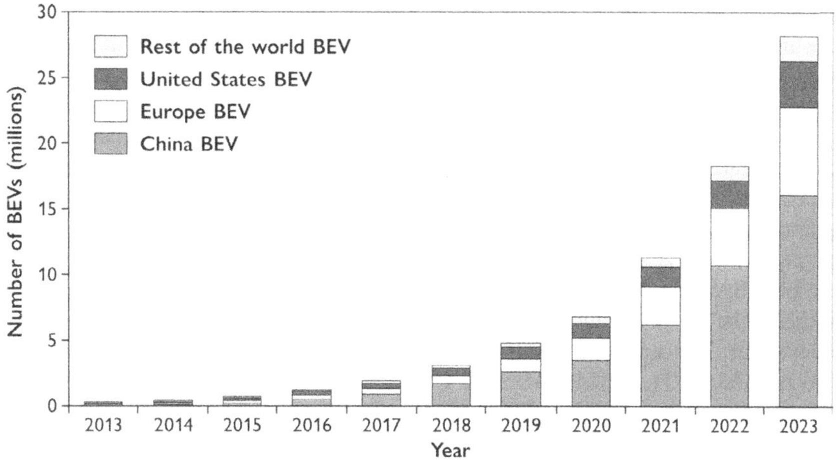

Fig. 5.1. Global Battery Electric Car Stock Trends, 2013–2023. *Source*: IEA (2024a); *Global electric car stock, 2013–2023*, IEA, Paris. https://www.iea.org/data-and-statistics/charts/global-electric-car-stock-2013-2023. Licence: CC BY 4.0. IEA analysis based on country submissions and data from ACEA. EAFO, EV Volumes and Marklines. *Notes*: BEV = battery electric vehicle. Includes passenger cars only.

pro-EV policy in a wealthy democracy and one known for harsh winter weather. From 1990, a package of EV incentives was introduced, including preferential tax and fee rates or complete exemptions. The motoring costs affected included car import tax, car purchase tax, car use taxes, road infrastructure tolls, and parking charges. Beyond financial measures, regulatory measures included the possibility for EV drivers to legally use bus lanes and the right (from 2017) for access to charging infrastructure in apartment buildings. From 2022, all public sector car purchases had to be EVs. By that year, BEVs had 79% market share, and fleet presence exceeded 20%. So, Norway already has a *fleet* share higher than most countries' *new car market* shares and is set to be the first country to become a 'zero-emission' new car market in 2025 (Norwegian EV Association, 2024). Norway's transition has been facilitated by its being a wealthy state. Ironically, its sovereign wealth fund is built on the taxation of fossil fuel production. However, the transition also required political leadership and social acceptance.

Elsewhere in Europe, despite most countries having committed to end ICEV sales by 2035, new vehicle market shares were measured in a handful of percentage points by 2020, while presence in established vehicle fleets remained negligible. Over the next four years, though, policies to incentivise and promote EVs began to have influence, and while the COVID-19 pandemic had impacts on the supply chains for all vehicles, the EV market actually grew in this period, for example, to around 15% of sales in Europe in 2022.

Nevertheless, by June 2024, concerns were expressed that the transition might be stagnating (Transport & Environment (T&E), 2024a). A core factor contributing to this concern was, to use the terminology of Diffusions of Innovation

by Volvo Cars has been criticised for understating the fuel consumption of the ICEV and the emissions associated with petrol production (Evans, 2023). Correcting these factors and allowing for the gradually decarbonising European electricity generation mix indicates a net contribution arising after 2-to-3 years.

However, a striking and ironic implication from a sustainable mobility perspective is that breakeven is brought forward by increasing annual distance travelled. A Volvo owner seeking to minimise car use, and so covering under 5,000 km per year, would need to own the vehicle for over five years before achieving carbon breakeven. The other side to this coin is that EVs are 'worth' more in carbon-reduction terms in parts of the world with relatively high annual distances travelled by car, such as the United States, where breakeven would be achieved relatively quickly. The same applies in the case of intensively used collective vehicles, such as taxis and potentially those in car-club or carpool fleets. Battery range, though, considered further below, represents a limit on some of those more intensive applications.

There are other factors to consider alongside carbon in judging the relative benefits of EVs over ICEVs, such as reducing noxious emissions, but such analyses underline how important it is to reduce EV production emissions if they are to contribute to NZ across the lifecycle. Battery technology options are one important consideration. Taking Volvo Cars (2024) again as the exemplar, its electric-only EX30 model for 2024 offered, conveniently for the purposes of comparative analysis, two battery options drawing on the two dominant technologies: a 51-kWh lithium iron phosphate (LPF) battery and a longer-range 69-kWh nickel, cobalt, and manganese (NMC) battery, which resulted in very similar vehicle specifications (e.g., in-service weights within 10 kg). However, LPF embodied emissions are typically around a third lower per kWh of capacity than NMC (IEA, 2024a). A complication in assessing more specific EV performance is that vehicle assemblers typically source batteries, the most important component in EV production, from supply chains, in contrast to the previous practice of making the most important components for ICEVs, engines and gearboxes, in-house. Whereas for ICEVs they had intimate knowledge about and control over key-component production, information about the GHG intensity of batteries has been harder to confirm. Nonetheless, based on an independently audited LCA, assuming 200,000 lifetime km, the EX30's LPF battery was assessed as contributing 15% to lifecycle emissions given European Union (EU) grid mix, whereas the NMC battery was expected to contribute nearly double, 28%, meaning the NMC model variant would have 22% higher lifecycle emissions overall (28 rather than 23 tCO_2eq), due to a combination of battery size and technology type. Here, the role of the consumer emerges as significant. The higher-range variant had a price premium of around 13%, delivering 42% greater range (480 rather than 340 km), but a user willing to recharge more often would have a personal car-use carbon footprint a fifth smaller.

The main ways that battery production emissions would be reduced is also dependent on technology. For NMH, the focus needs to be on decarbonising the material inputs and for LPF decarbonising the electricity supplied for manufacture in producer states, notably China. IEA (2024a) identifies a pathway to reduce embedded emissions by 35% by 2035. GHG performance must also be linked to the wider sustainability of EV production. While detailed discussion of these important issues is beyond the scope of this chapter, initiatives to inform producer and consumer decision-making are emerging. For example, the Global Battery Alliance's (2024) 'Battery Passport' rates batteries on a range of environmental, social, and governance criteria covering the battery supply chain, including raw material extraction. Among other factors, the criteria cover environmental degradation beyond GHG emissions, the use of child and forced labour, the rights of workers and indigenous communities, and instances of corruption in the governance processes.

Last in this section, which has provided a commentary on EVs largely comparative with ICEVs, it is important to observe that ICEVs may currently have lower embedded carbon, but they do not present a relevant benchmark for NZ auto production. Both ICEVs and EVs remain dependent on carbon-intensive steel, on average around 650 kg per new bodyshell (Hill et al., 2024). For an EV steel accounts for an estimated 16%–27% of embedded emissions. The pathway to decarbonisation is in part to use steel produced using renewable hydrogen and electricity, while in part substituting other materials for steel (T&E, 2024c). Again, consumers can play a role by choosing smaller vehicles which use less steel, although the recent context has been a trend for vehicle size and weight to increase (T&E, 2024d).

In summary, while much has been made of EVs potentially being emissions-free in use, there is a long way to go before we have a genuinely (net)-zero-emission vehicle, and for this and other reasons, the image of the EV as an 'environmentally friendly' vehicle is subject to major caveats. Nonetheless, EVs offer the potential to be NZ in a way that other technologies do not rival. The next section considers how far the potentially-NZ EV appeals to the consumer.

CHALLENGE 2: INSERTING THE ELECTRIC CAR INTO AUTOMOBILITY

The society–car nexus is an interrelated set of challenges linking current travel and social practices, vehicle technology, and the sector of the economy which provides for vehicle use. Under the regime of 'automobility' (Sheller & Urry, 2000), the majority of households in the wealthy democracies are now dependent upon the car, in the sense that removing access would leave them excluded from opportunities and needs or at least disadvantaged with respect to those with cars. Sociotechnical explanations are key to understanding the transition (Geels, 2012). According to that perspective, the EV represents a niche technology that must compete with a regime that has its origins in the early 20th century and whose domination has intensified over the decades since the 1950s. Hence, the ICEV represents over 100 years of collective financial, political, and emotional investment. Research and development have increased the reliability, capability, and ease of use of the car, even if myths exaggerate the car's advantages, and downplay the time and money costs invested in ownership (Wolf, 1996).

As a result, travel patterns, spatial development, and the provision of goods and services have co-evolved with mass adoption of the car, such that they are hard to disaggregate (Freund & Martin, 2000). Facilities have been centralised and relocated to places with lower land cost based on the possibility of access by private car. Traditional public transport networks either do not provide an access option within reasonable walking range or that option is complex, slower, and less reliable than using a car. The scheduling of travel requirements through the day has also become more complex, involving multi-purpose, multi-stage trips, and the auto-mobile society has established expectations about the spontaneity of trips, which limits the possibility to plan journeys around fixed route and timetable public transport or 'active travel' (Pooley et al., 2011), with women with caring responsibilities particularly affected (Jain et al., 2011; see Chapter 16). While the reversal of development trends to deliver forms of urbanisation that instead favour travel patterns not dependent on the car has been strongly recommended as a policy objective (Banister, 2011), even in national contexts with a strong tradition of regulatory spatial planning such as The Netherlands, policy interventions to reduce car-dependent travel patterns have shown only partial success, due to the strength of the regime (Schwanen et al., 2004).

The societal and economic relations underpinned by the car reflect its capabilities or 'level of service'. Liquid fuels are highly energy dense. 'Recharging' an ICEV to deliver a 750-km range takes less than five minutes. Another key aspect of performance is that the auto industry has honed production of new ICEVs to a price point that a wide range of consumers can bear. The best-selling new car in Europe in the first half of 2024, the Dacia Sandero, a 'B segment' (small-to-medium size) vehicle, also included in its range the lowest-cost 'Essential' model, available from €11,990 in France (Crooks, 2024). Even the lowest-specification version (Groupe Renault, 2024) provided five seats and driving performance within the lower end of the range for this class, as well as several driver-assistance and safety technologies required by European legislation to be standard equipment. Most strikingly, the Sandero's consumption measured by the standard Worldwide Harmonised Light Vehicles Test Procedure (WLTP) was 5.4l/100 km. Combined with a 50-l petrol tank, the implied potential range is 935 km.

It is into this sociotechnical context that the EV makes its entrance …

EV technology is less complex mechanically than ICEV technology, the fundamentals of the electromagnetic motor having changed little across a heritage longer than the ICE. This simplicity gives EV technology several consumer advantages, particularly to users who are not emotionally invested in the ICEV 'driver experience'. ICEV driving has traditionally been technically 'performative' (Edensor, 2004), such as through the timing of gear changes to road context and engine noise, even if in the last decade cars have increasingly become automated and managed, as manufacturers complied with lower emission regulations. EVs represent a further change to clutch-less gear-free performance, eliciting different driver behaviours (Labeye et al., 2016) due to features such as quiet operation and 'one-pedal driving', which replaces the application of friction brakes with energy-regenerative braking on release of the accelerator. Electric technology integrates particularly well with driver-assistance technologies, and potentially, full automation (Chapter 6).

EV complexity instead tends to rise with digital engineering, for example, to optimise battery charging in a range of ambient temperatures. The EV is in the vanguard of the march towards the 'software-defined' (Juliussen, 2021) car: 'over the air' software updates to implement enhancements based on learning-in-use are a particular feature of marques like Tesla, reducing the need for physical service centres. Little or no engine maintenance, the absence of a gearbox and clutch, and lower wear on the service brakes all reduce the costs of ownership and can be expected to reduce the carbon emissions associated with the servicing supply chain. One area where service costs may be higher is in respect of tyres, which may currently be considered specialist. Concerns have also been raised about the environmental impacts of EV tyres wearing faster, due to a combination of higher weight and higher potential rates of acceleration than an equivalent ICE car. Actual tyre wear would depend on tyre specification, and higher weight could be compensated for by driving style, although drivers may well be tempted to use the available acceleration (Emissions Analytics, 2024). On balance, though, considering the car lifespan measure of purchase and operating costs – the 'total cost of ownership' (TCO) – EVs have been identified as the lowest-cost technology option in many cases, a situation expected to expand across car market segments (Element Energy, 2021).

Nonetheless, despite the identification of consumer benefits, barriers to consumer acceptance remain, both real and constructed. Concern about limited knowledge or misunderstanding of EVs led European consumer organisation BEUC (2022) to address consumer concerns in a briefing on climate impact and the durability of EVs and their batteries. FairCharge (2024), a consortium of UK motor sector organisations, went further, feeling it necessary to produce a consumer-oriented publication to address no fewer than 43 'myths' it had identified as reducing the acceptance of EVs. The e-booklet considered a wide-ranging set of issues. At one extreme were real differences between EVs and ICEVs but which were seen to be exaggerated. At the other, some of the claims were seen to be entirely invented, possibly to encourage engagement with digital media to enhance advertising revenues ('clickbait'). Populist politics has also been identified as a source of poorly evidenced negative narratives about EVs, linking them to the wider argument that policies to reduce environmental impacts were unreasonably limiting individuals' freedom to choose the lifestyles and products they wanted. For example, in the United Kingdom in September 2023, such claims reached the highpoint of official sanction, with the then prime minister, announcing the removal of 'unreasonable environmental impositions' on the British people, including delaying the limit on sales of new ICE-only cars from 2030 to 2035 (UK Government, 2023).

Focusing on the more evidenced barriers to consumer acceptance, and the actual technical competitiveness of the EV with the ICEV, the important issues tend to lead back to the traction battery: the size, weight, and cost per unit capacity and its output and recharging characteristics. A kilogram of petrol (approximately 1.4 l) powers an ICEV for around 25 km. Considering current commercialised technology, a kilogram of lithium-ion battery moves a typical EV around 1 km. Hence, adding EV range adds significantly to vehicle weight and size, which in turn reduces efficiency. Moreover, how much range 1 kg of battery offers reduces according to conditions such as cold weather, wind, vehicle load, use of auxiliary equipment, carrying or towing loads outside of the vehicle

comparison is an extreme one but serves to underline the point that auto manufacturers have struggled to approach price and configuration parity, in particular at the low-cost end of the market. Expensive batteries represent a higher proportion of the cost for low-price vehicles and on high-value vehicles can also be compensated for in terms of consumer appeal by novel features such as Tesla's 'autopilot' driver-assistance system. However, the industry in Europe has also been criticised for focusing on profits from EVs rather than sales volume by 'gaming' regulatory targets (T&E, 2024a).

In the United Kingdom, relatively high pricing meant that just 9% of private buyers of new cars chose an EV in July 2024, compared with 22% of corporate buyers (although the latter represent over 60% of the market). A price-compensating factor for companies was that the vehicles could be provided or leased to employees so that they benefited from tax incentives, while the employer would be able to zero-rate commuting by EV in its mandatory 'Scope 3' carbon emission declaration. Nonetheless, by November 2024, year-to-date sales for all purchaser types gave a BEV market share of 18.7% (Society of Motor Manufacturers and Traders (SMMT), 2024), and it was clear the sector would not achieve the first annual market share target (22%) of the zero-emission mandate (UK Government, 2024). Given that the targets for subsequent years were to rise sharply to 80% by 2030, the industry's own view was that more government incentives to offset purchase and charging costs were necessary (Autotrader, 2024). One such measure that has been applied for car-dependent private consumers on low and medium incomes in France is 'social leasing' of EVs (T&E, 2024e).

Fig. 5.3 indicates the EV price premia for five national markets in 2018 and 2022. The situation varied between car segments and between states, but overall, the four wealthy democracies saw declines between 2018 and 2022, although still showing differentials of 14%–59%. China in contrast showed a particularly rapid reduction, leading to a large negative price differential in the small vehicle segment, seen by the IEA to result from the mature EV market, and providing support for the view that EV-ICEV cost parity will be achieved globally soon.

The signs were in 2024 that this aspect of the transition is evolving. EV battery packs were found to account for 40% of the vehicle cost in 2020, although expected to fall to 25% by 2030 (König et al., 2021). The supply-chain cost to an auto manufacturer of lithium-ion battery packs has fallen sharply in recent years, as low as $100/kWh (€90) in April 2024 (IEA, 2024b), implying the cost of the 50-kWh array typical of a high production volume EV would cost €5,000 (giving an WLTP range of around 400 km). Several mainstream manufacturers were committing to launching bespoke EV models in the €20–25,000 price range, spurred on by the need to meet regulated sales targets (T&E, 2024a). For example, Stellantis Group (Vauxhall Motors Ltd, 2024) made much of launching its Frontera model in the United Kingdom with both EV and hybrid powertrains offered at a matched price of £23,495 (€28,000).

Related to the issue of relatively high purchase price was that of fast depreciation, already identified by 2021 as meaning that first owners might not experience a TCO benefit even if subsequent owners would (Element Energy, 2021). Depreciation affects all cars but is particularly severe in the first three years of life, when value might fall by 40%–60%, after which the rate of depreciation slows.

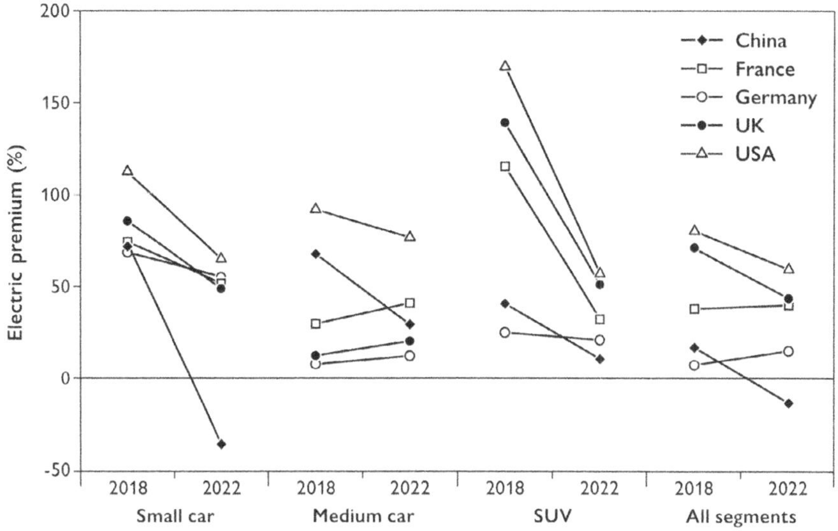

Fig. 5.3. Price Gap Between the Sales-weighted Average Price of Conventional and Electric Cars in Selected Countries, Before Subsidy, by Size, in 2018 and 2022. *Source*: IEA (2024a); https://www.iea.org/data-and-statistics/charts/price-gap-between-the-sales-weighted-average-price-of-conventional-and-electric-cars-in-selected-countries-before-subsidy-by-size-in-2018-and-2022. License: CC BY 4.0.

There is evidence that depreciation has been overall higher for EVs (IEA, 2024a). There are several reasons why this might be. One is a lack of consumer confidence in the residual value of the battery and how to measure that. Another is that the reduction in costs of new EVs referred to above has, to an extent, undercut the value of the used market. The technology is also undergoing ongoing rapid evolution. Battery capacity for equivalent models has progressively increased. The first EV to achieve significant sales globally was the Nissan Leaf. Battery capacity has grown in four increments since 2011 from 24 to 62 kWh, more than doubling the range from 160 to 340 km. Range per unit cost is likely to be an influence on many purchasers, pushing down the value of used vehicles with limited performance. Therefore, depreciation might support sales volumes in the used market but risks deterring purchasers of new vehicles.

A third EV cost concern for consumers has been insurance, which has been higher for EVs due to underwriters pricing in uncertainties about the risks around the new technology. The principal uncertainties included the severity of fires resulting from collisions, and a lack of capability to ascertain if a battery has been damaged following a collision and whether it can be repaired satisfactorily and economically, or must be 'written off' (Rafferty, 2022). In response to price differentials in the United States, Tesla introduced in-house insurance services in 2019, indicating it could save customers up to 30%, because it could more accurately price the risks relating to its own products (Tesla, 2019). In the United Kingdom, some underwriters suspended offering EV insurance in 2023, while the premium differential was found to be 10%–25% higher according to the model. The lack of skilled repair technicians and problems with spare part logistics were identified as additional EV-specific explanatory factors (FairCharge, 2024).

The development of EV charging networks at a sufficient pace to support the transition has also been identified as a key requirement, with both quantity and range of charging speeds being important (IEA, 2024a). Early adoption of EVs has been closely linked to having access to private charging facilities in Norway, which had achieved a vehicles-to-home charger ratio approaching 1:1 by 2023 and a vehicles-to-public charger ratio of 24:1 (Guttman, 2023). However, China represents a different model. It has one of the lowest rates of home charging (IEA, 2024a) but a relatively high ratio of vehicles-to-charger of 7:1 (Zheng & Zhuang, 2023) and has installed 60% of the world's slow-speed chargers and over 90% of its fast chargers (IEA, 2024a) (Fig. 5.4). However, vehicles-to-charger ratios are

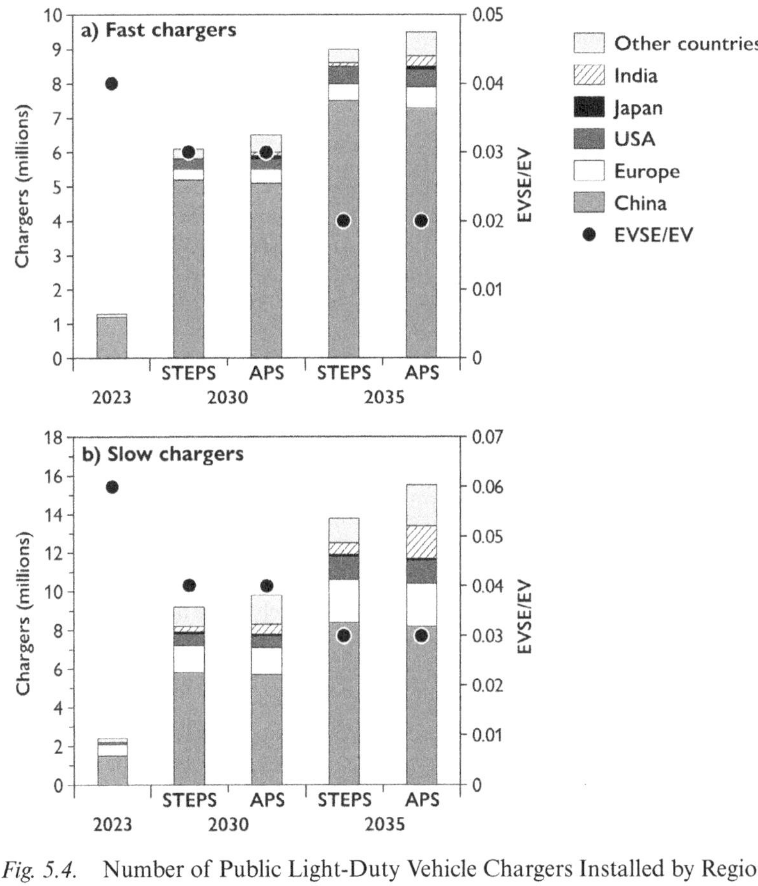

Fig. 5.4. Number of Public Light-Duty Vehicle Chargers Installed by Region, 2023–2035. *Source*: IEA (2024a); https://www.iea.org/data-and-statistics/charts/number-of-public-light-duty-vehicle-fast-chargers-installed-by-region-2023-2035 and https://www.iea.org/data-and-statistics/charts/number-of-public-light-duty-vehicle-slow-chargers-installed-by-region-2023-2035. License: CC BY 4.0. *Notes*: RoW = rest of world; STEPS = stated policies scenario; APS = announced pledges scenario; EV = electric vehicle; EVSE = electric vehicle supply equipment.

widening as the EV fleet grows and consumers do report difficulties linked to the distribution, maintenance, and speed of chargers (Zheng & Zhuang, 2023).

Other states must decide where they fit into these broad strategies emphasising private vs public facilities. In response to a survey in the United Kingdom in 2024, 81% of EV owners reported that they mainly charged at home or at their places of employment, and this was supported by an overall vehicles-to-charger ratio of 1.2:1, a ratio much lower than for public chargers (17.0:1) (ChargeUK, 2024). Some private solutions for those without off-street parking have been implemented by connecting a cable from the domestic supply under or over the public realm to the EV, but these still tend to rely on the individual being able to park on the public highway adjacent to the home. For this reason, off-street parking is often considered a necessity for home charging, but it was estimated that 37% of all UK households did not have off-street parking (ChargeUK, 2024).

It remains a point of debate how far home charging is a prerequisite for EV ownership. A survey of 25,000 EV drivers in six European states found that 44% reported not having a home charger (Shell Recharge, 2023) confirming that for some users, it is not essential. Similarly, a preference survey of 1,200 Indians, none of whom were recorded as yet owning an EV, found that two-thirds would not be deterred by a lack of home charging, although 55% of respondents said they would have access to home charging, and another 30% thought they could make changes to facilitate it (McKinsey, 2023). However, the IEA (2024a) reports on the commitments of the wealthy democracies to provide public charge points by 2035. Given the scenarios for EV fleet growth, the predicted ratios will actually increase to 50–80 vehicles per public charger. Whether charger availability reduces in practice depends on interactions with factors such as future frequency and speed of charging events, but the estimates suggest the de facto approach is closer to Norway's reliance on home charging than the Chinese strategy, leaving questions as to how those without access to off-street parking will be able to charge easily. One radical alternative which has been applied in niches around the world but is not currently deployed for light-duty vehicles in Europe or North America is battery-swap technology, which functions by holding a stock of charged batteries at a swap station, with the depleted battery swapped for a charged one in a time equivalent to refuelling with liquids (IEA, 2024a).

Without such solutions, the politics of private vs public charging in the European states with many homes without off-street parking are likely to become more important as the market moves beyond early adopters. One reason concerns the relative cost of the solutions (Parkhurst & Clayton, 2022). Home charging is undertaken at domestic rates. In liberalised electricity markets, there are generous incentives for off-peak charging when there is excess supply over demand, so EV charging is often at below marginal cost. The growth of photovoltaic panel installation means that some charging in the summer may be at zero cost (although it may be at the expense of foregoing export earnings). In contrast, public EV charging in the wealthy democracies is now a commercial service. Installation costs may be subsidised, but residual capital, operational, energy costs, and profit margin all need to be covered from user charges. There may be a tax differential: in the United Kingdom, value-added tax on domestic energy is at a special rate of 5%, but the general rate of 20% is applied to energy purchased through public

Corradi, C., Sica, E., & Morone, P. (2023). What drives electric vehicle adoption? Insights from a systematic review on European transport actors and behaviours. *Energy Research & Social Science*, *95*, 102908, https://doi.org/10.1016/j.erss.2022.102908

Crooks, A. (2024). *The Dacia Sandero is Europe's best selling car, so far*. Auto Express 18 July 2024. Retrieved August 31, 2024, from https://www.autoexpress.co.uk/dacia/sandero/363817/dacia-sandero-europes-best-selling-car-so-far

Davey, J. A. (2007). Older people and transport: Coping without a car. *Ageing & Society*, *27*(1), 49–65.

Edensor, T. (2004). Automobility and national identity: Representation, geography and driving practice. *Theory, Culture & Society*, *21*(4–5), 101–120. https://doi.org/10.1177/0263276404046063

Element Energy. (2021). *Electric cars: Calculating the total cost of ownership for consumers*. Final report for BEUC. Retrieved September 1, 2024, from https://www.beuc.eu/sites/default/files/publications/beuc-x-2021-039_electric_cars_calculating_the_total_cost_of_ownership_for_consumers.pdf

Emissions Analytics. (2024). *Why tyres kill fish*. Podcast 11 May 2024. Retrieved September 1, 2024, from https://www.emissionsanalytics.com/podcast/2024/5/11/why-tyres-kill-fish-mamps-episode-03

Evans, S. (2023). Factcheck: 21 misleading myths about electric vehicles. Retrieved May 27, 2025, from https://www.carbonbrief.org/factcheck-21-misleading-myths-about-electric-vehicles/

FairCharge. (2024). *The little book of EV myths*. Retrieved September 1, 2024, from https://www.faircharge.co.uk/little-book-of-ev-myths

Fevang, E., Figenbaum, E., Fridstrøm, L., Halse, A. H., Hauge, K. E., Johansen, B. G., & Raaum, O. (2021). Who goes electric? The anatomy of electric car ownership in Norway. *Transportation Research Part D: Transport and Environment*, *92*, 102727. https://doi.org/10.1016/j.trd.2021.102727

Freund, P., & Martin, G. (2000). Driving south: The globalization of auto consumption and its social organization of space. *Capitalism Nature Socialism*, *11*(4), 51–71. https://doi.org/10.1080/10455750009358940

Geels, F. W. (2012). A socio-technical analysis of low-carbon transitions: Introducing the multi-level perspective into transport studies. *Journal of Transport Geography*, *24*, 471–482. https://doi.org/10.1016/j.jtrangeo.2012.01.021

Global Battery Alliance. (2024). *Our work: Battery passport*. Retrieved August 31, 2024, from https://www.globalbattery.org/battery-passport/

Groupe Renault. (2024). *Dimensions & spécifications: Dacia Sandero*. Retrieved September 1, 2024, from https://www.dacia.fr/gamme-vehicules/sandero/dimensions-specifications.html

Guttman, B. (2023, August 15). *Norway still leads the way for EV sales, but must now focus on its charging infrastructure*. Roland Berger. Retrieved September 4, 2024, from https://www.rolandberger.com/en/Insights/Publications/EV-Charging-Index-Expert-insight-from-Norway.html

He, L. (2024, April 24). *A brutal elimination round is reshaping the world's biggest market for electric cars*. CNN. Retrieved September 3, 2024, from https://edition.cnn.com/2024/04/24/business/china-ev-industry-competition-analysis-intl-hnk/index.html

Hill, N., Scammell, H., Raugei, M., Kilstein, A., & King, A. (2024). *The use of green steel in the automotive industry*. Ricardo AEA Ltd Report ref. ED18758 for Transport & Environment. Retrieved August 31, 2024, from https://www.transportenvironment.org/articles/cleaning-up-steel-in-cars-why-and-how

International Energy Agency. (2024a). *Global EV outlook 2024*. Retrieved August 31, 2024, from https://www.iea.org/reports/global-ev-outlook-2024

International Energy Agency. (2024b). *Batteries and secure energy transitions*. World Energy Outlook Special Report. Retrieved September 4, 2024, from https://www.iea.org/reports/batteries-and-secure-energy-transitions

International Panel on Climate Change. (2022). Summary for policymakers. In P. R. Shukla, J. Skea, R. Slade, A. Al Khourdajie, R. van Diemen, D. McCollum, M. Pathak, S. Some, P. Vyas, R. Fradera, M. Belkacemi, A. Hasija, G. Lisboa, S. Luz, & J. Malley (Eds.), *Climate change 2022: Mitigation of climate change* (pp. 1–48). Contribution of Working Group III to the Sixth Assessment Report of the Intergovernmental Panel on Climate Change. Cambridge University Press. https://doi.org/10.1017/9781009157926.001

Jain, J., Line, T., & Lyons, G. (2011). A troublesome transport challenge? Working round the school run. *Journal of Transport Geography*, *19*(6), 1608–1615. https://doi.org/10.1016/j.jtrangeo.2011.04.007

Jaramillo, P., Kahn Ribeiro, S., Newman, P., Dhar, S., Diemuodeke, O. E., Kajino, T., Lee, D. S., Nugroho, S. B., Ou, X., Hammer Strømman, A., & Whitehead, J. (2022). Transport. In P. R. Shukla, J. Skea, R. Slade, A. Al Khourdajie, R. van Diemen, D. McCollum, M. Pathak, S.

Some, P. Vyas, R. Fradera, M. Belkacemi, A. Hasija, G. Lisboa, S. Luz, & J. Malley (Eds.), *IPCC, 2022: Climate change 2022: Mitigation of climate change* (pp. 1049–1160). Contribution of Working Group III to the Sixth Assessment Report of the Intergovernmental Panel on Climate Change. Cambridge University Press. https://doi.org/10.1017/9781009157926.012

Juliussen, E. (2021). *How software-defined cars build on connected-car concepts*. Retrieved December 11, 2024, from https://www.embedded.com/how-software-defined-cars-build-on-connected-car-concepts/

König, A., Nicoletti, L., Schröder, D., Wolff, S., Waclaw, A., & Lienkamp, M. (2021). An overview of parameter and cost for battery electric vehicles. *World Electric Vehicle Journal*, *12*(1), 21. https://doi.org/10.3390/wevj12010021

Labeye, E., Hugot, M., Brusque, C., & Regan, M. A. (2016). The electric vehicle: A new driving experience involving specific skills and rules. *Transportation Research Part F: Traffic Psychology and Behaviour*, *37*, 27–40. https://doi.org/10.1016/j.trf.2015.11.008

McKinsey. (2023, September 14). *Consumers are driving the transition to electric cars in India*. Retrieved September 3, 2024, from https://www.mckinsey.com/industries/automotive-and-assembly/our-insights/consumers-are-driving-the-transition-to-electric-cars-in-india#/

Motor. (2022, January 24). *Engine and NAF's large range test: So much range was lost in the winter cold.* Retrieved September 2, 2024, from https://www.motor.no/aktuelt/motors-store-vintertest-av-rekkevidde-pa-elbiler/217132 (automatically translated from Norwegian).

Newman, P., & Kenworthy, J. (2011). 'Peak car use': Understanding the demise of automobile dependence. *World Transport Policy & Practice*, *17*(2), 31–42.

Norwegian EV Association. (2024). *Norwegian EV policy*. Retrieved September 3, 2024, from https://elbil.no/english/norwegian-ev-policy/

Parkhurst, G., & Clayton, W. (2022). Conclusion: The electric car as a component of future sustainable mobility. Chapter 12. In G. Parkhurst & W. Clayton (Eds.), *Electrifying mobility: Realising a sustainable future for the car* (pp. 231–246). Emerald. https://doi.org/10.1108/S2044-994120220000015002

Pooley, C. G., Horton, D., Scheldeman, G., Tight, M., Jones, T., Chisholm, A., Harwatt, H., & Jopson, A. (2011). Household decision-making for everyday travel: A case study of walking and cycling in Lancaster (UK). *Journal of Transport Geography*, *19*(6), 1601–1607. https://doi.org/10.1016/j.jtrangeo.2011.03.010

Rafferty, I. (2022). Underwriters 'don't understand the risks' of EV accidents. *Insurance Times*, May 4. Retrieved September 4, 2024, from https://www.insurancetimes.co.uk/analysis/underwriters-dont-understand-the-risks-of-ev-accidents/1441012.article?adredir=1

Renault Italia. (2024). *Nuova spring*. Renault Italia S.P.A. Retrieved September 2, 2024, from https://www.dacia.it/gamma-ibrida-ed-elettrica/spring-city-car.html

Rogers, E. M. (2003). *Diffusions of innovations* (5th ed.). Free Press.

Schwanen, T., Dijst, M., & Dieleman, F. M. (2004). Policies for urban form and their impact on travel: The Netherlands experience. *Urban Studies*, *41*(3), 579–603. https://doi.org/10.1080/0042098042000178690

Sheller, M., & Urry, J. (2000). The city and the car. *International Journal of Urban and Regional Research*, *24*, 737–757. https://doi.org/10.1111/1468-2427.00276

Shell Recharge. (2023). *EV driver survey report 2023*. Retrieved September 4, 2024, from https://shell-recharge.com/en-gb/solutions/knowledge-centre/reports-and-case-studies/ev-driver-survey-report#dawn-of-a-new-electric-era

Society of Motor Manufacturers and Traders. (2024). *Car registrations*. Retrieved December 12, 2024, from https://www.smmt.co.uk/vehicle-data/car-registrations/

Tesla. (2019, August 28). *Introducing Tesla insurance*. Retrieved September 4, 2024, from https://www.tesla.com/en_gb/blog/introducing-tesla-insurance

Transport & Environment. (2024a). Stagnation and growth: The European EV market How EU regulation and carmaker strategies impact the EV market. *Briefing*, June 2024. Retrieved September 2, 2024, from https://www.transportenvironment.org/uploads/files/2024_06_EV_market_briefing.pdf

Transport & Environment. (2024b). Bridging the gap: Carmakers' progress toward the 2025 car CO_2 targets. *Briefing*, April 2024. Retrieved September 2, 2024, from https://www.transportenvironment.org/articles/bridging-the-gap-carmakers-progress-toward-the-2025-car-co2-targets/

Transport & Environment. (2024c). Cleaning up steel in cars: Why and how? *Briefing*, July 2024. Retrieved August 31, 2024, from https://www.transportenvironment.org/articles/cleaning-up-steel-in-cars-why-and-how

Transport & Environment. (2024d, January). *Ever-wider: Why large SUVs don't fit, and what to do about it. The case for European law-makers to review the maximum width of light duty vehicles.* Retrieved December 11, 2024, from https://www.transportenvironment.org/articles/ever-wider-why-large-suvs-dont-fit-and-what-to-do-about-it

Transport & Environment. (2024e, November). *Social leasing: A key measure for national social climate plans.* Retrieved December 11, 2024, from https://www.transportenvironment.org/articles/social-leasing-a-key-measure-for-national-social-climate-plans

UK Government. (2023, September 20). *Prime Minister Rishi Sunak sets out his new approach to net zero* (speech). Retrieved September 1, 2024, from https://www.gov.uk/government/speeches/pm-speech-on-net-zero-20-september-2023

UK Government. (2024). *Pathway for zero emission vehicle transition by 2035 becomes law.* Retrieved December 12, 2024, from https://www.gov.uk/government/news/pathway-for-zero-emission-vehicle-transition-by-2035-becomes-law

Vauxhall Motors Ltd. (2024). *New Vauxhall Frontera electric & petrol hybrid both starting from £23,495.* Retrieved December 12, 2024, from https://www.vauxhall.co.uk/cars/frontera/overview.html

Volvo Cars. (2020). *Carbon footprint report battery electric XC40 recharge and the XC40 ICE.* Retrieved August 31, 2024, from https://www.volvocars.com/images/v/-/media/applications/pdpspecificationpage/my24/xc40-electric/pdp/volvo-cars-LCA-report-xc40.pdf

Volvo Cars. (2024). *Carbon footprint report EX30.* Retrieved August 31, 2024, from https://www.media.volvocars.com/global/en-gb/media/pressreleases/325272/volvo-ex30-lca-shows-lowest-carbon-footprint-of-any-fully-electric-volvo-car-to-date

Wolf, W. (1996). *Car mania. A critical history of transport.* Pluto Press.

Yang, Z. (2023). How did China come to dominate the world of electric cars? *MIT Technology Review*, February 21. Retrieved September 3, 2024, from https://www.technologyreview.com/2023/02/21/1068880/how-did-china-dominate-electric-cars-policy/

Zheng, R., & Zhuang, J. (2023, August 15). *EV sales continue to rise in China, but its charging infrastructure is struggling to keep up.* Roland Berger. Retrieved September 4, 2024, from https://www.rolandberger.com/en/Insights/Publications/EV-Charging-Index-Expert-insight-from-China.html

CHAPTER 6

OPPORTUNITIES AND CHALLENGES IN HARNESSING THE NET ZERO IMPACTS OF AUTONOMOUS VEHICLES

Eda Beyazit

University of the West of England, UK

ABSTRACT

Many nations have set net zero targets to lower emissions, yet the transport sector is falling short. Autonomous vehicles (AVs) could significantly decrease carbon emissions, especially when they use renewable energy sources and adopt shared public transport. However, these beneficial outcomes are intertwined with various factors and assumptions. Based on a systematic literature review, this chapter examines AVs' capacity to reduce carbon emissions while also addressing issues such as the potential rise in vehicle kilometres travelled (VKTs) and barriers to adoption. It underscores the need for electrification, renewable energy use, and widespread AV adoption to mitigate environmental effects. This chapter assesses how AV technology could influence the journey towards net zero objectives and highlights possible challenges, opportunities, and burdens associated with the shift towards AVs.

Keywords: Autonomous vehicles; connected and autonomous vehicles; decarbonisation of transport; net zero; renewable energy

Towards Transport Net Zero
Transport and Sustainability, Volume 20, 83–100

Published under exclusive licence by Emerald Publishing Limited
ISSN: 2044-9941/doi:10.1108/S2044-994120250000020006

energy sources (Vaidya & Mouftah, 2020). On the other hand, AVs may increase VKTs (Annema, 2020; Pan et al., 2021; Taiebat, Brown, et al., 2019; Taiebat, Stolper, et al., 2019) due to new user groups (youth, older populations, and individuals with disabilities) and new travel behaviours (Bhardwaj et al., 2023). According to Litman (2023), the most notable effects – such as decreased traffic and parking congestion, enhanced independent mobility for low-income individuals (resulting in a lowered demand for public transit), improved safety, energy savings, and reduced pollution – will likely only become substantial once AVs are widespread and economically accessible, anticipated between the 2040s and 2060s.

Most promisingly in the context of this book, however, electrification, powered by the use of renewable energy sources (Gawron et al., 2019; Jones & Leibowicz, 2019), and a 100% penetration of AVs could significantly reduce transport's environmental impacts (Conlon & Lin, 2019; Patella et al., 2019; Sinha et al., 2021). At high penetration rates, lowering driver unpredictability can decrease road congestion and minimise wasted energy (Milakis et al., 2017). The potential is even more significant if AVs are used in shared modes rather than as privately owned automated vehicles (Jones & Leibowicz, 2019; Pfaffenbichler et al., 2020). Shared mobility, particularly full automation in public transport, offers several benefits, such as enhancing the availability and quality of services by improving frequency, operating hours, and coverage (Beyazit et al., 2024). It also elevates convenience and comfort through a mix of scheduled and on-demand services. These improvements can significantly benefit disadvantaged individuals and communities by eliminating the exclusionary character of AVs arising from issues such as financial unaffordability and physical and digital accessibility.

Focusing on urban and rural passenger mobility, this chapter proceeds by applying a systematic literature review to (1) understand how automation and connected transport systems impact decarbonisation, (2) explore adoption barriers in various contexts for AVs, and (3) identify potential challenges, opportunities, and burdens during the transition to AVs. It concludes with a discussion of a future research agenda.

METHODOLOGY

Due to the multidisciplinary nature of AV research, a holistic approach was undertaken to avoid excluding any relevant topics. Among the widely recognised search databases such as Google Scholar, Web of Science, and Scopus, the last of these was selected due to its more comprehensive database. The search query included all possible variations of automated cars in titles, abstracts, and keywords. These included 'autonomous vehicle$', 'autonomous car$', 'automated vehicle$', 'automated car$', 'autonomous driving', 'automated driving', 'driver*less vehicle$', 'driver*less car$', 'self*driving vehicle$', 'self*driving car$', 'robo*taxi', 'autopilot vehicle$' and 'autopilot cars'. Since net zero research is also highly interdisciplinary, the search query included extensive disciplinary coverage including engineering, computing, mathematics, social sciences, earth

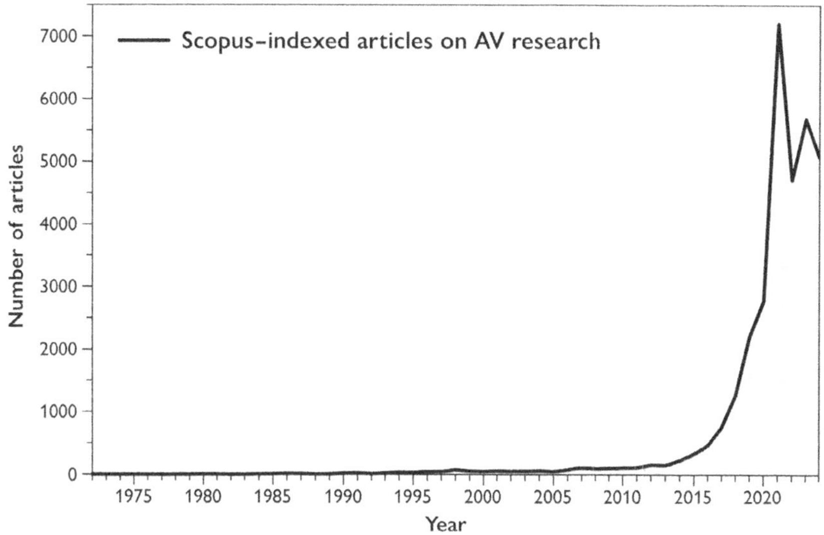

Fig. 6.1. Scopus-indexed Articles on AV Research Between 1972 and 2024. *Source*: Author.

sciences, energy, economics, psychology, physics, arts, and health. The search was limited to the English language and to journal papers only and concluded with 32,405 documents produced between 1972 and 2024 (Fig. 6.1).

Autonomous transport has certainly attracted significant attention in academic research, especially in the last decade. It is unsurprising that the smallest number of articles, 439, were produced between 1972 and 2000. This number increased to 838 articles between 2000 and 2010. The number of academic outputs reached over 1,000 in 2007, following a linear trend until 2014. From 2014 onwards, there was an exponential growth, peaking at 7,202 articles in 2021.

The visualisation software VOSviewer was used to extract the key terms used in the articles, and a co-occurrence analysis of keywords used in article titles and abstracts was conducted to visualise the conceptual landscape of the research on AVs since 1972. A total of 284,120 keywords was extracted. Each term is counted only once in a document, and the minimum occurrence of a term is set to five. Out of 16,572 terms that met the threshold, 9,943 terms (60%) were used to calculate the relevance scores. The terms were arranged into seven clusters and visualised by the VOSviewer tool (Fig. 6.2). These clusters are related to (1) the control and optimisation of AVs, trajectory tracking, and topics related to underwater AVs; (2) connected AVs, platooning, traffic efficiency, and lane management; (3) motion sensors and global positioning; (4) imaging, light detection and ranging (lidar), and trajectory prediction; (5) internet of things, blockchain, and algorithms; (6) participant/respondent perceptions, trust, and travel costs; and (7) themes related to methods and databases. Based on this macro analysis of the emerging themes within AV research, a noticeable gap in the discussion on the carbon reduction or net zero impacts of AVs is evident.

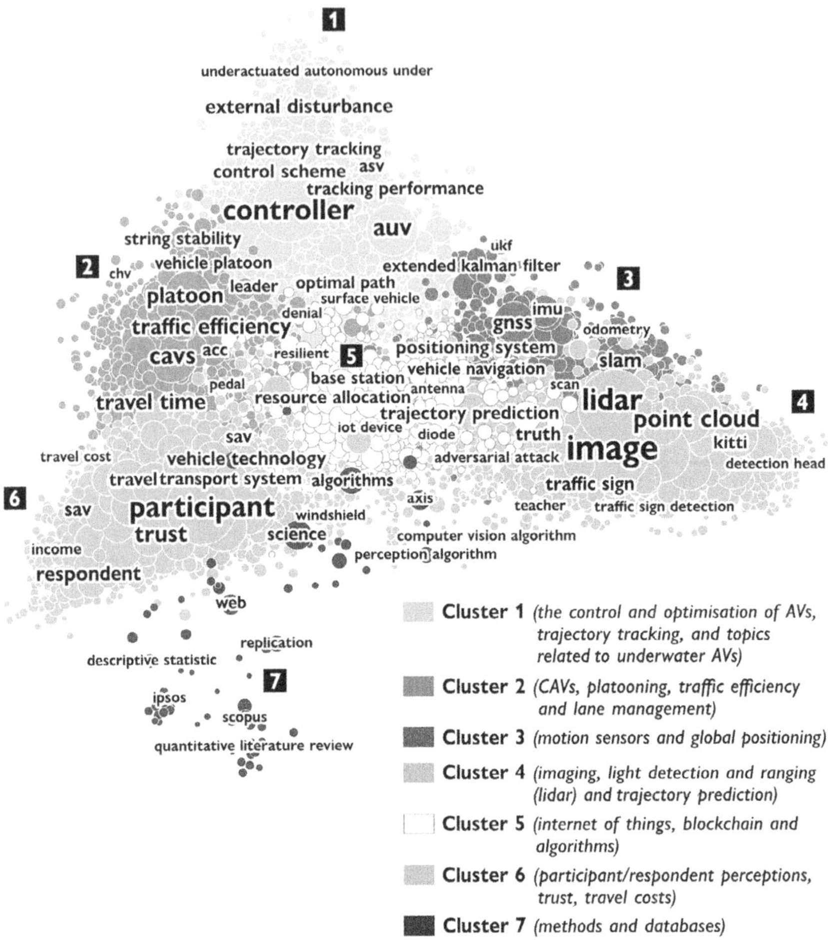

Fig. 6.2. Key Terms and Clusters in AV Research (1972–2024). *Source*: Author.

Since the primary objective of this study is to examine the potential of AVs in helping to achieve net zero targets, an extensive literature review was carried out, focusing on the most relevant research within this domain. The search was limited to the years 2014–2024 and yielded a total of 30,671 papers. Subsequently, a secondary search using authors' keywords was conducted, with a focus on keywords such as carbon, CO_2, net zero, and climate, resulting in the identification of 66 relevant papers. After reviewing the abstracts of 66 papers, 46 of them did not pertain to the transport planning and policy domains (for example, they were related to meteorology, microstructures, photodetectors, materials, mining, and underwater structures), or focused mainly on logistics (some of which are acknowledged in this chapter), and one had only an accessible abstract. Two additional papers were eliminated after a brief review of the complete texts, as they did not have a clear AV focus.

A further six papers provided literature reviews or utilised qualitative data, such as interviews with stakeholders, which have been used to guide the contextual discussions but were not included in an in-depth review. For instance, in a discussion of academic and grey literature on the European context, Grindsted et al. (2022) refer to the sustainability impacts of AVs from a policy, planning, and governance perspective. Li and Miao (2023) present a comprehensive literature review on the potential environmental impacts of AV technology while discussing the legal and policy framework of China's carbon neutrality goals. Greenwald and Kornhauser (2019) explore the climate implications of AVs in relation to travel behaviour and the economic implications of AVs, such as reductions in accidents, travel time, and increases in sales, and discuss how these can be translated into policy in the context of the United States.

Ultimately, a total of 11 papers featuring original case-study research findings and quantifying the net zero impacts of AVs were selected for a thorough review. Before going on to discuss these, it should be noted that systematic reviews have limitations. It can be challenging to find relevant articles if their abstracts, titles, and keywords do not contain predefined concepts. Moreover, limiting the research to one or two search engines may fail to capture valuable research published on other platforms. However, the results indicate the variety of discussions provided by the selected papers, and this chapter reports on their findings, offering an international perspective on the potential positive and negative effects of AVs on transport decarbonisation objectives.

THE IMPACTS OF AUTOMATED AND CONNECTED VEHICLES ON DECARBONISING TRANSPORT

The research on AVs' carbon reduction impacts shows some divergence, and achieving actual reductions will need additional policies and various trade-offs. AVs can affect GHG emissions in various ways, such as changes in driving efficiency, vehicle size, traffic congestion, service type (private or shared), fuel type, charging systems, and travel demand for both current and new users, as well as new habits encouraged by having the autonomous driving ability that may change the VKT (e.g., Bhardwaj et al., 2023; Jones & Leibowicz, 2019). This section presents the key factors associated with the deployment of AVs and their impacts on carbon emissions.

Three Revolutions: Electrification, Shared Systems, and AVs

Together, the electrification of vehicles, shared mobility, and autonomous technologies have the potential to transform the future of transport (Fulton et al., 2017; Sperling, 2018). In this sense, researchers find that the electrification of AVs is essential for their potential carbon reduction impacts (Huang et al., 2024), as is deploying shared AVs (SAVs), which should be accompanied by policies that discourage conventional personal car use (Cremer et al., 2021). As Roca-Puigròs et al. (2023) stress, the dearth of research into the potential impact of carbon reduction can be addressed by integrating the three renowned revolutions: fleet electrification, shared mobility, and vehicle automation.

Predicting a green cost model for autonomous battery electric vehicles (ABEVs), human-driven EVs (HDEVs) and human-driven fossil-fuelled vehicles (HDFV), Li and Chen (2023) created three simulation models – the accuracy of the EV energy consumption, mixed traffic flow, and intersection signalling. The authors conclude that EVs are more suitable for urban commuting as their green cost is less than that of fuel vehicles under different speeds, acceleration, and deceleration conditions. They have also demonstrated that as the ABEVs' penetration into the mixed traffic increases, the average green cost decreases compared to a scenario with HDFVs.

Utilising an agent-based simulation model (SUMO), Huang et al. (2024) produced a traffic flow model in Suzhou, China. According to the authors, introducing a few CAVs into predominantly human-driven traffic causes human drivers to behave unpredictably, negating the benefits of CAVs. Similar to the findings of Fan et al. (2024), the authors also suggest that as AVs become more common and gain permission to operate on public roads (ADPR), controlling vehicle speed, route planning, and distance between cars improves where the vehicles are gasoline powered (GV). Their calculations show that EVs can reduce carbon emissions by 70%–90% compared to GVs on urban roads during peak hours, while CAVs can further reduce emissions by 35%–50% compared to HDVs. This progress gradually cuts carbon emissions: 'CAV technology has a limited impact on vehicle carbon emissions at low ADPRs, and its advantages can be realized at high ADPRs' (Huang et al., 2024, p. 7).

Zong and Yue (2023) conducted several simulation experiments to examine how various disturbance characteristics affect carbon emissions in mixed traffic of CAVs and HDVs with internal combustion engines. They explored how the positions of CAVs and the diversity of human factors in mixed traffic influence carbon emissions amid different disturbances. They introduced the concept of a disturbance impact coefficient, which can be utilised by CAVs to mitigate carbon emissions in mixed traffic, particularly under various disturbance regimes caused by other CAVs or HDVs. According to their experiments, each CAV introduced to mixed traffic can impact five HDVs. While the nearest HDV might generate higher carbon emissions, the total emissions would be reduced. Additionally, they recommended establishing a CAV-only lane when the proportion of CAVs in traffic surpasses 16.7%. This is in line with Wadud et al.'s (2016) argument that without human interaction with the road surface, therefore, without the need for speed controls, the AVs could reach higher speeds, boosting energy efficiency by 7%–22%. In terms of reducing the adverse impacts of human behaviour, such as delays in reacting to changes, slowing down, and potential accidents, Zong and Yue (2023) also suggested implementing a maximum speed limit of 17.5 m/s (about 60 km/h) in road sections near intersections, entrance ramps, and exit ramps to reduce variability in human behaviour.

As noted earlier, CAVs may demand significant energy because they depend on onboard and roadside sensing, communication, and computing devices. These components can consume considerable energy due to added weight, computing requirements, sensor demands, and increased drag (Qu et al., 2022). Although simulation models showed optimistic results, Qu et al. (2022) emphasised the necessity of a real-world example. They developed four simulation models of a chosen road in China, including intersections, signalling, and mixed traffic. The HDV and AV models were based on field trials conducted on the road, whereas CAV and cooperative CAV (CCAV) scenarios stemmed from computer simulations. They discovered that AVs alone lead to higher energy consumption because of the additional sensing, communication and computing devices, slow response to infrastructure control units, and cautious car-following settings with more stops along the route, leading to longer travel time. In contrast, the CAV and CCAV scenarios show improved energy consumption compared to HDV and AV scenarios. The authors highlight the importance of developing CV technology to improve AVs, making them more compatible with human drivers' behaviour and enhancing their future energy performance.

Other Aspects of AVs

In addition to the enhanced driving performance of AVs compared to HVs, Bhardwaj et al. (2023) also highlight some technological changes that may play a role in carbon reduction. AVs are projected to enhance safety over conventional cars by minimising accidents. This improvement might permit eliminating safety features like airbags, thus reducing vehicle weight and cutting energy intensity by 7%–10% (Taiebat, Brown, et al., 2019). Lightweight vehicles might motivate consumers to switch to smaller personal cars, due to enhanced safety and better accommodation of specific travel needs. With automated car-sharing and on-demand mobility models, aligning vehicle size with the travel group's size can optimise the distance-weighted average fuel economy, potentially reducing energy intensity by 21% (Wadud et al., 2016). This might also encourage consumers to buy/use smaller cars for individual trips, decreasing energy intensity, captured by lower ASCs for smaller AVs.

On the downside, evidence from the literature shows that AVs will likely encourage VKTs. First, autonomous driving features may attract new users to the market, including older people (see Chapter 17), individuals with limited mobility, and those (including children) without driving licences. Evidence from the United States and Canada indicates a potential increase in private car owners as a result of a transition to AVs, leading to a rise in VKT ranging from 14% to 57% (Bhardwaj et al., 2023). Bhardwaj et al. (2023) estimate that the availability of AVs will increase new users, potentially leading to a 20% rise in carbon emissions. Looking at the existing trends in the United States in 2009, Wadud et al. (2016) estimated a 2%–10% increase in overall personal vehicle travel from new users. Second, lower fuel costs and higher efficiency may lead to increased driving. VKT may also be triggered by high parking costs, prompting a habit of returning the vehicle instead of paying high parking fees (Bhardwaj et al., 2023). Moreover, Wadud et al. (2016) estimate that, for SAVs, deadheading (travelling empty to pick up passengers) could boost the VKT by up to 10%, thereby negating the advantages gained from decreased demand for personal vehicles.

Bhardwaj et al. (2023) investigate consumer behaviour comprehensively by including in their study endogenous factors such as the prices introduced by the automakers, focusing on the deployment of privately owned vehicles. Considering three policy scenarios – Canada's baseline policies, baseline plus tax, and baseline plus zero-emission vehicles (ZEVs) – along with optimistic and pessimistic views on AV uptake, the authors acknowledge that the AV adoption rate in the country is likely to be uncertain. In terms of consumer behaviour, they gather that under strong policies that support ZEVs, the uptake of AVs would be slower compared to a situation where strong ZEV policies are not present. They reveal that due to Canadians' reluctance to pay more for AVs, they would be more likely to switch to AVs with combustion engines. Therefore, the rate of future uptake of AVs will depend upon the interplay between consumers' willingness to pay and purchase costs of AVs.

DISCUSSION: CHALLENGES, OPPORTUNITIES, AND BURDENS OF AV ADOPTION

This review has illustrated that AVs have the potential to significantly lower carbon emissions from road transport. Various simulations show potential emission reductions ranging from 50% to over 95% (e.g., Fan et al., 2024; Huang et al., 2024; Roca-Puigròs et al., 2023). Yet, in all the models reviewed, certain assumptions and factors that underlie the emissions reduction calculations come into play. The literature highlights fleet electrification, charging strategies, high market penetration rates, vehicle technology, size and weight, and ownership costs, among other factors (Table 6.2).

Electrification of the AV vehicle fleet is crucial for achieving greater reductions in carbon emissions. The majority of authors refer to the importance of electrification rates even if they do not directly include EVs in their models. Ridesharing and SAVs demonstrate higher environmental benefits due to reduced VKT despite new user groups adopting AV technology (Dong et al., 2022). The advancement of AVs equipped with connected vehicle technology has further potential to reduce carbon emissions, improve traffic management, and promote eco-friendly driving (Fan et al., 2024; Qu et al., 2022; Zong & Yue, 2023). Another crucial element is the market penetration rate: integrating AVs into mixed traffic can greatly lower GHG emissions, particularly when the traffic is composed entirely of AVs (e.g., Li & Chen, 2023; Zong & Yue, 2023).

However, several factors may diminish these benefits. The most prominent negative factor is the potential increase in VKT due to AVs attracting new users (Bhardwaj et al., 2023; Wadud et al., 2016). Second, lower fuel costs and high parking fees can increase VKT due to more empty runs (Wadud et al., 2016). Third, even with the progress in green energy, the energy consumption of AVs due to onboard and roadside sensing, communication, and computing devices should be considered a potential adverse factor (Qu et al., 2022). Finally, high purchase costs of AVs may put off consumers and reduce market penetration (Bhardwaj et al., 2023). Therefore, real-world deployment of AV or SAV may not achieve full penetration instantly but will rely significantly on long-term consumer adoption and behaviour.

Future adoption of new and emerging technologies may require specific policy support. To enhance the potential benefits of reduction, it is essential to implement policies that promote clean energy, including a carbon tax, initiatives for zero-emission vehicles, a shift to renewable energy sources, and the regional organisation of independent power grids, among others. Moreover, researchers believe policymakers need to prioritise incentives that encourage automobile manufacturers to invest in developing advanced technologies and establish laws and regulations that support advanced AV technology, including vehicle-to-vehicle (V2V) and vehicle-to-everything (V2X) communications (Wadud et al., 2016). However, even after the AV fleet has completely decarbonised, concerns regarding possible rises in VKT should be acknowledged. Thus, rather than concentrating on private vehicles, efforts should prioritise promoting autonomous public transport. This strategy, combined with concerted efforts to promote active mobility, would accelerate the achievement of net zero goals.

CONCLUDING REMARKS: DO WE NEED AN AV FUTURE TO REACH NET ZERO?

The future of AVs remains uncertain. Numerous factors, from regulations to technological advancements, must align to achieve a net zero future driven by AVs. In the meantime, transport planners and policymakers are equipped with various tools to reduce carbon emissions in transport. These include strategies discussed throughout this *Transport and Sustainability* book series, such as increasing public transport usage, encouraging walking, cycling, and wheeling, electrifying mobility, and promoting equity in transport.

A significant portion of the literature on AVs originates from both advanced and emerging economies. The studies examined in this review include case analyses from China, the United States, Switzerland, and Canada. In fact, the United States and China are the leading countries that dominate the AV intellectual landscape, followed by Germany, the United Kingdom, South Korea, Australia, the Netherlands, France, Japan, Singapore, Spain, Italy, and Sweden (Sadeghpour & Beyazit, 2024). Although the research field extends beyond these countries to include developing regions (e.g., da Silva et al., 2022; Sovacool et al., 2022), these are encountered less frequently. Despite the United States and China driving the highest pollution levels in the transport sector (Statista, 2023), thereby justifying increased volumes of research into emerging technologies, achieving net zero goals remains a global target. All countries should be given the opportunity and the capacity to develop their unique policy, planning, and technology solutions. International research funding and collaboration agreements are beneficial in this regard.

While the opportunities in AV technology and market integration are nascent and may require time to fully develop, the rollout of AVs/CAVs could significantly help achieve net zero targets, particularly in cities with high traffic volumes. Three revolutions (shared, autonomous, and electric), combined with connected transport, result in the highest predicted reduction of energy consumption and GHG emissions. Promoting AVs and fostering technological advancement can involve diversifying their applications, including options like micro-transit services for business parks and late-night transport, first and last-mile connectivity, airport shuttles, access to transport hubs, and integration with traditional public transit systems (Greenwald & Kornhauser, 2019). Introducing AVs and CAVs in controlled settings will enhance our comprehension of their benefits and challenges. This approach can lead to steady advancements in technology and interactions with other users, drivers, and the surrounding environment, ultimately fostering a broader public appeal and addressing concerns arising from equity dimensions of AV deployment (Beyazit et al., 2024). In this sense, transport policy needs to make stronger connections with spatial planning practices as well as social sciences.

In uncertain futures, collaboration among various stakeholders is vital. The private sector and research fields cannot independently lead the development of technology and the deployment of AVs/CAVs. Governments, along with their associated departments and resources, must establish an appropriate regulatory and economic environment to foster collaboration. Ultimately, the potential impacts of AV on achieving net zero are both multidisciplinary and involve multiple stakeholders. Given collaborative environments are essential for enhancing net zero outcomes in any other domain, future advancements in AV technology aimed at reaching net zero could also positively influence other areas of technology and transport-related policy.

REFERENCES

Annema, J. A. (2020). Policy implications of the potential carbon dioxide (CO_2) emission and energy impacts of highly automated vehicles. *Advances in Transport Policy and Planning*, *5*, 149–162.

Beyazit, E., Soh, E.. & Martens, K. (2024). Rescuing transport from inequities: How can ACT contribute to a more inclusive transport system? In M. Attard, Y. Shiftan, & N. Thomopoulos (Eds.), *Sustainable automated and connected transport* (pp. 139–159). Emerald. https://doi.org/10.1108/S2044-9941202419

Bhardwaj, C., Axsen, J., & Crawford, C. (2023). Simulating long-term emissions from private automated vehicles under climate policies. *Transportation Research Part D: Transport and Environment*, *118*, 103665.

Conlon, J., & Lin, J. (2019). Greenhouse gas emission impact of autonomous vehicle introduction in an urban network. *Transportation Research Record: Journal of the Transportation Research Board*, *2673*(5), 142–152.

Cremer, A., Müller, K., & Finkbeiner, M. (2021). A systemic view of future mobility scenario impacts on and their implications for city organizational LCA: The case of autonomous driving in Vienna. *Sustainability*, *14*(1), 1–19. https://doi.org/10.3390/SU14010158

da Silva, T. B., Baptista, P., Santos Silva, C. A., & Santos, L. (2022) Assessment of decarbonization alternatives for passenger transportation in Rio de Janeiro, Brazil. *Transportation Research Part D Transport and Environment*, *103*, 103161.

Department for Business, Energy and Industrial Strategy. (2021). *UK territorial greenhouse gas emissions national statistics*. BEIS.

Department for Transport. (2021). *Decarbonising transport*. https://assets.publishing.service.gov.uk/media/610d63ffe90e0706d92fa282/decarbonising-transport-a-better-greener-britain.pdf

Dong, J., Li, Y., Li, W., & Liu, S. (2022). CO_2 emission reduction potential of road transport to achieve carbon neutrality in China. *Sustainability*, *14*(9), 5454.

Ercan, T., Onat, N. C., Keya, N., Tatari, O., Eluru, N., & Kucukvar, M. (2022). Autonomous electric vehicles can reduce carbon emissions and air pollution in cities. *Transportation Research Part D: Transport and Environment*, *112*, 103472.

Fan, R., Lu, J., Wang, D. Z. W., & Zhang, F. (2024). Adaptive speed optimization strategy at signalized intersections based on the penetration rate of connected automated vehicles. *IEEE Transactions on Intelligent Transportation Systems*, *25*(10), 13122–13135.

Fulton, L., Mason, J., & Meroux, D. (2017). *Three revolutions in urban transportation*. UC Davis Institute for Transportation & Development Policy. https://itspubs.ucdavis.edu/publication_detail.php?id=2723

Gawron, J. H., Keoleian, G. A., De Kleine, R., Wallington, T. J., & Kim, H. C. (2019). Deep decarbonization from electrified autonomous taxi fleets: Life cycle assessment and case study in Austin, TX. *Transportation Research Part D-Transport and Environment*, *73*, 130–141.

Greenwald, J. M., & Kornhauser, A. (2019). It's up to us: Policies to improve climate outcomes from automated vehicles. *Energy Policy*, *127*, 445–451. https://doi.org/10.1016/j.enpol.2018.12.017

Grindsted, T. S., Christensen, T. H., Freudendal-Pedersen, M., Friis, F., & Hartmann-Petersen, K. (2022). The urban governance of autonomous vehicles – In love with AVs or critical sustainability risks to future mobility transitions. *Cities*, *120*, Article 103504.

Huang, K., Zhou, P., Liu, Z., Tang, T., Zhang, H., & Jiang, W. (2024). The calculation and distribution of CAV carbon emissions on urban transportation systems: A comparative analysis of renewable and non-renewable energy sources. *Renewable Energy*, *230*, 120884.

International Energy Agency. (2023). *Tracking clean energy progress 2023*. https://www.iea.org/reports/tracking-clean-energy-progress-2023

International Panel on Climate Change. (2023). *Climate change 2023 synthesis report*. https://www.ipcc.ch/report/ar6/syr/downloads/report/IPCC_AR6_SYR_LongerReport.pdf

Jones, E. C., & Leibowicz, B. D. (2019). Contributions of shared autonomous vehicles to climate change mitigation. *Transportation Research Part D: Transport and Environment*, *72*, 279–298.

Li, C., & Chen, Y. (2023). Evaluation and comparative analysis of the mixed traffic flow on urban roads considering the green cost. *Sustainable Energy Technologies and Assessments*, *57*, 103292.

Li, X.-W., & Miao, H.-Z. (2023). How to incorporate autonomous vehicles into the carbon neutrality framework of China: Legal and policy perspectives. *Sustainability*, *15*(7), 5671.

Li, Y., Li, X., & Jenn, A. (2022). Evaluating the emission benefits of shared autonomous electric vehicle fleets: A case study in California. *Applied Energy*, *323*, Article 119638.

Litman, T. (2023). *Autonomous vehicle implementation predictions: Implications for transport planning*. Victoria Transport Institute. https://www.vtpi.org/avip.pdf

Milakis, D., Van Arem, B., & Van Wee, B. (2017). Policy and society related implications of automated driving: A review of literature and directions for future research. *Journal of Intelligent Transportation Systems*, *21*(4), 324–348.

Pan, S., Fulton, L., Roy, A., Jung, J., Choi, Y., & Gao, H. O. (2021). Shared use of electric autonomous vehicles: Air quality and health impacts of future mobility in the United States. *Renewable Sustainable Energy Reviews*, *149*, 111380. https://doi.org/10.1016/J.RSER.2021.111380

Parkhurst, G., & Seedhouse, A. (2019). Will the 'smart mobility' revolution matter? In I. Docherty & J. Shaw (Eds.), *Transport matters* (pp. 349–380). Policy Press.

Patella, S. M., Scrucca, F., Asdrubali, F., & Carrese, S. (2019). Carbon footprint of autonomous vehicles at the urban mobility system level: A traffic simulation-based approach. *Transportation Research Part D:Transport and Environment*, *74*, 189–200.

Pfaffenbichler, P., Gühnemann, A., Klementschitz, R., Emberger, G., & Shepherd, S. (2020). A systemic analysis of impacts of individual and shared automated mobility in Austria. *Transportation Research Procedia*, *49*, 6–16.

Poynting, M. (2024, February 8). *World's first year-long breach of key 1.5C warming limit*. BBC News. https://www.bbc.co.uk/news/science-environment-68110310

Qu, X., Zhong, L., Zeng, Z., Tu, H., & Li, X. (2022). Automation and connectivity of electric vehicles: Energy boon or bane? *Cell Reports Physical Science*, *3*(8), 101002.

Roca-Puigròs, M., Marmy, C., Wäger, P., & Beat Müller, D. (2023). Modeling the transition toward a zero emission car fleet: Integrating electrification, shared mobility, and automation. *Transportation Research Part D: Transport and Environment*, *115*, 103576.

Sadeghpour, M., & Beyazıt, E. (2024). Exploring the landscape of autonomous vehicles research: A scientometric analysis in the context of urban transportation planning. *Journal of Scientometric Research, 13*(1), 25–42.

SAE. (2021). *Taxonomy and definitions for terms related to driving automation systems for on-road motor vehicles*. https://www.sae.org/standards/content/j3016_202104/

Sinha, A., Bassil, D., Chand, S., Virdi, N., & Dixit, V. (2021). Impact of connected automated buses in a mixed fleet scenario with connected automated cars. *IEEE Transactions on Intelligent Transportation Systems, 23*(8), 11982–11993.

Sovacool, B. K., Daniels, C., & AbdulRafiu, A. (2022). Transitioning to electrified, automated and shared mobility in an African context: A comparative review of Johannesburg, Kigali, Lagos and Nairobi. *Journal of Transport Geography, 98*, 103256.

Sperling, D. (2018). *Three revolutions steering automated, shared, and electric vehicles to a better future* (1st ed.). Island Press.

Statista. (2023). *Greenhouse gas emissions of the transportation sector worldwide in 2023*. https://www.statista.com/statistics/1084166/ghg-emissions-transportation-sector-globally-by-country/#:~:text=The%20United%20States%20is%20the,China%2C%20India%2C%20and%20Russia.

Taiebat, M., Brown, A., Safford, H. R., Qu, S., & Xu, M. (2019). A review on energy, environmental, and sustainability implications of connected and automated vehicles. *Environmental Science & Technology, 52*(20), 11449–11465.

Taiebat, M., Stolper, S., & Xu, M. (2019). Forecasting the impact of connected and automated vehicles on energy use: A microeconomic study of induced travel and energy rebound. *Applied Energy, 247*, 297–308.

Vaidya, B., & Mouftah, H. (2020). Connected autonomous electric vehicles as enablers for low-carbon future. *Research Trends and Challenges in Smart Grids*. https://doi.org/10.5772/intechopen.84287.

Wadud, Z., MacKenzie, D., & Leiby, P. (2016). Help or hindrance? The travel, energy and carbon impacts of highly automated vehicles. *Transportation Research Part A: Policy and Practice, 86*, 1–18.

Wang, Y., Wang, Y., Lin, Q., Yin, M., Yin, M., Liu, J., Chen, W., Jin, L., & Han, Y. (2023). Study on the start-up operation and sludge characteristics of organic wastewater treated by the EGSB–two-stage anaerobic/aerobic biofilm process. *Desalination and Water Treatment, 290*, 1–11.

Zong, F., & Yue, S. (2023). Carbon emission impacts of longitudinal disturbance on low-penetration connected automated vehicle environments. *Transportation Research Part D: Transport and Environment, 123*, 103911.

CHAPTER 7

EIGHT PARKING REFORMS TO IMPROVE CITIES AND COOL THE PLANET

Donald Shoup

UCLA, USA

ABSTRACT

This chapter proposes eight parking reforms to improve cities and cool the planet: (1) require solar panels on parking lots, (2) convert garages into housing, (3) make parking fines progressive, (4) give discounts at parking meters for residents and cleaner cars, (5) offer on-street parking permits that allow residents to block their own driveways, (6) convert private parking into public parking, (7) unbundle parking from housing, and (8) cash-out employer-paid parking. These parking reforms are simple, cheap, and fast ways to reduce carbon emissions.

Keywords: Parking; sustainability; equity; urban design; parking cash-out; solar panels

INTRODUCTION

The Way to Multiply Big Problems Is to Neglect Small Ones[1]

Cars are parked 95% of their lives, so the rubber meets the road mainly in parking spaces. Parked cars don't produce any carbon emissions, so parking reforms might seem to be a poor way to reduce carbon emissions. Nevertheless, parking reforms are the simplest, cheapest, and quickest way to reduce carbon emissions from transportation.

Towards Transport Net Zero
Transport and Sustainability, Volume 20, 101–114

Published under exclusive licence by Emerald Publishing Limited
ISSN: 2044-9941/doi:10.1108/S2044-994120250000020007

Fig. 7.2. A Rendered Conversion of a Garage into an Apartment in Los Angeles, Before and After. *Source*: Author.

Design review can ensure that a garage conversion is consistent with the house and the neighbourhood. The most prominent feature of a new house is often a two-car garage door as big as a billboard, and a garage apartment can improve the house's design (Fig. 7.2). Cars can still park in the driveway. An affordable housing opportunity has been hiding in plain sight for decades: allow property owners to convert their garages into apartments.

MAKE PARKING FINES PROGRESSIVE

Drivers who don't pay at meters reduce the funds available for city services. Drivers who stay longer than the time limit make it harder for other drivers to find a space. Double parking blocks a whole lane of traffic. Illegal parking in a disabled space makes life even more difficult for people with disabilities. Parking violations have victims and parking enforcement is necessary.

Modest fines are enough to convince most drivers to obey parking regulations but not repeat offenders who view the fines as an acceptable gamble or just another cost of doing business. For example, Court (2022) reported on one London millionaire who bragged that he parks wherever he wants because he doesn't worry about paying the fines. 'Broke people see a fine, but I see VIP parking'.

Progressive parking fines can deter frequent violators without unfairly punishing anyone else (Shoup, 2018). In Claremont, CA, for example, the first ticket for overtime parking in a calendar year is \$50, the second \$100, and the third \$150. Stronger punishments for serial offenders are a standard legal practice, and they can also apply to parking scofflaws. Serial offences will be expensive (Shoup, 2010).

Progressive fines were impossible when enforcement officers didn't know how many previous tickets a car had received. Ticket-writing devices now connect to the city's database and automatically assign the fine according to the number of previous tickets for the licence plate. For minor violations like overtime parking at a meter, some cities issue a warning for the first offence and progressively higher fines for subsequent ones. The warnings show the city aims for compliance, not just revenue. Only repeat offenders pay progressively higher fines.

Progressive parking fines make it politically easier for cities to enforce parking regulations because many tickets can be warnings, not fines. If cities want to collect more money from parking tickets, they can increase the fines for chronic offenders without penalising anyone else.

GIVE DISCOUNTS AT PARKING METERS FOR RESIDENTS AND CLEANER CARS

Cities can charge market rate prices to manage curb parking and provide public revenue, but drivers still want to park for free. How can cities generate political support for market prices? Some cities create political support for parking meters by giving residents a discount. In Miami Beach, Florida, non-residents pay $4 an hour at meters, but residents pay only $1. Monterey, CA, gives residents the first two hours free in municipal lots and garages.

Pay-by-licence-plate technology enables these resident discounts. Drivers pay by mobile phone or at a parking kiosk, and cars with licence plates registered in a city automatically receive discounts (Shoup, 2014, 2018). Like hotel taxes that generate local revenue without unduly burdening the residents, parking meters with resident discounts are the least objectionable to the greatest number of voters. They resemble Monty Python's proposal to solve Britain's economic problems by taxing foreigners living abroad.

Cities can also give discounts for cars with lower exhaust emissions. For example, parking meters in Madrid charge 20% less for low-emission cars and 20% more for high-emission cars. According to the head of Madrid's sustainability division, 'We thought it would be fair if the cars that pollute more pay more and compensate those who use more efficient vehicles' (Kassam, 2014). Lyon charges parking prices in proportion to the weight of the parked vehicles. Paris charges non-residents €18 an hour for parking SUVs in the city centre, triple the price residents pay (Chrisafis, 2024). Meters can also give discounts for shorter cars that take up less curb space. If meters give discounts for shorter and cleaner cars, more people will drive them. To avoid an undue burden on low-income families with older cars, cities can also give parking discounts to low-income drivers.

Residents will see several advantages of market rate curb parking with discounts for residents and for cleaner, smaller cars. First, residents will pay discounted prices. Second, they will quickly find open curb spaces. Third, they will face less traffic congestion. Fourth, they will breathe cleaner air. Fifth, the meter revenue will pay for added public services.

Parking discounts for residents and smaller, cleaner cars may seem complicated, but few voters will object to them. The discounts for residents will create political support for parking meters, and the environmental discounts for smaller, cleaner cars will benefit everyone.

OFFER ON-STREET PARKING PERMITS THAT ALLOW RESIDENTS TO BLOCK THEIR OWN DRIVEWAYS

Every driveway displaces a curb parking space. To solve this problem, California authorises cities to issue permits that allow residents to park in the curb lane in front of their own driveways. These block-your-own-driveway permits give residents guaranteed parking spaces in front of their homes. To allow this, Section 22507.2 of the California Vehicle Code says,

> A local authority may, by ordinance, authorize the owner or lessee of property to park a vehicle in front of the owner's or lessee's private driveway when the vehicle displays a permit issued pursuant to the ordinance authorizing such parking.

The curb lane in front of every home's driveway becomes a reserved parking space for the residents. Because California and most other states prohibit parking in front of a driveway, driveway parking permits do not take on-street spaces away from anyone else. By allowing residents to park in front of their driveways, the driveway permits will reduce the demand for the other curb spaces and thereby make on-street parking more available for everyone else.

Selling block-your-own-driveway permits might sound like privatising public streets, but cities now give curb cuts to property owners without any charge. Cities can instead charge for driveway permits. For example, Hermosa Beach, CA, charges $170 for a driveway permit that is good until the property is sold or the tenant moves out.

In densely populated areas where residents greatly outnumber the curb cuts, cities could go further and charge property owners for all curb cuts (Jonlin, 2023). If the curb parking revenue pays for public services that most residents want, charging property owners for curb cuts that already privatise public space seems reasonable.

If cities offer to rent reserved curb spaces to property owners, residents may want to pay for a reserved curb space rather than build a garage with a driveway and a curb cut. Building a garage, a driveway, and a curb cut will suddenly become much more expensive than paying for a reserved parking space on the street.

Reserved curb spaces are especially appropriate in historic neighbourhoods where owners sometimes insert an incongruous garage into the front of an old house because it is the only way to guarantee a parking space. As a result, the garages deface historic buildings, undermine the pedestrian environment, and reduce the available curb parking space (Shoup, 2005a, pp. 457–460). If owners can rent a curb space without building a garage, they will save money and have more living space. The city will receive more revenue and preserve its historic fabric.

CONVERT PRIVATE PARKING INTO PUBLIC PARKING

Businesses have discovered a new way to earn revenue from the off-street parking spaces they provide free for their customers and employees. To take advantage of the spaces not used by their employees and customers, the property owners contract with private parking operators to make the unused spaces available to the general public, at a price (Shoup, 2018). The property owner and the parking operator share the resulting revenue.

The simplest way to convert a free private lot into paid parking is to reserve the spaces for a business when it is open and make the spaces available to the paying public when the business is closed (Fig. 7.3). The parking operator manages the lot and shares the revenue from otherwise empty spaces with the property owner. Private parking can also be converted into public parking all day, with validations for employees and customers.

Many cities manage their own public parking lots, and they can also manage private parking lots. In Sacramento, CA, the city's Parking Services division manages 24 private parking lots with 4,810 spaces and splits the meter revenue with the property owners. The city receives all the fine revenue for meter violations in the lots. In 2023, the 24 lots earned $3 million ($624 per space), and the fine revenue was $424,000 ($88 per space). Sacramento effectively added 24 public parking lots at no cost to the city.

Because private operators cannot issue tickets for meter violations, they have trouble enforcing the drivers' responsibility to pay in private lots except by booting or towing, which are expensive and unpopular. To solve the enforcement problem, Sacramento has another programme: private operators manage the lots and the city enforces the meters. In 2023, Sacramento earned $112,000 in ticket revenue for enforcing the meters in 20 private lots with 1,237 spaces ($93 per space). City enforcement enables private operators to manage public parking.

Fig. 7.3. Private Parking Converted to Public Parking. *Source*: Author.

value, and an equivalent contribution to another fringe benefit such as better health insurance. A policy will fail the test only if it subsidises parking more than it subsidises the alternatives to solo driving.

California's cash-out requirement does not tax or prohibit employer-paid parking. Instead, cash-out gives commuters a new choice, rewards the alternatives to solo driving, reduces vehicle trips, enhances employee welfare, treats all commuters equally, costs employers little or nothing, and sidesteps the political opposition to charging for parking at work. It achieves all these goals by letting commuters choose how to spend their own earnings.

In 2020, Washington, DC, adopted the Transportation Benefits Equity Amendment, modelled on California's cash-out law. Employers who have 20 or more employees and rent spaces to subsidise parking for drivers must offer an equivalent benefit to employees who commute without a car.

To encourage parking cash-out for rented parking spaces, Chapter 9.28.110 of the Santa Monica, CA, Municipal Code requires separate leases for parking in non-residential property: 'All off-street parking spaces associated with new non-residential projects ... shall be leased or sold separately from the rental or purchase fees'. This leasing arrangement will make it easy for employers to offer cash in lieu of a free parking space.

How does parking cash-out change commuters' travel choices? To answer this question, the California Air Resources Board sponsored research on the results of offering cash-out (Shoup, 2005b, 2018). After employers offered cash-out, solo driving to work fell by 17%, carpooling increased by 64%, transit ridership increased by 50%, and walking or biking increased by 39% (Fig. 7.4).

Consider a representative group of 100 commuters. Before cash-out, 76 drove to work alone. Cash-out induced 13 of these solo drivers to shift to another mode: nine joined carpools, three began to ride transit, and one began to walk or bike to

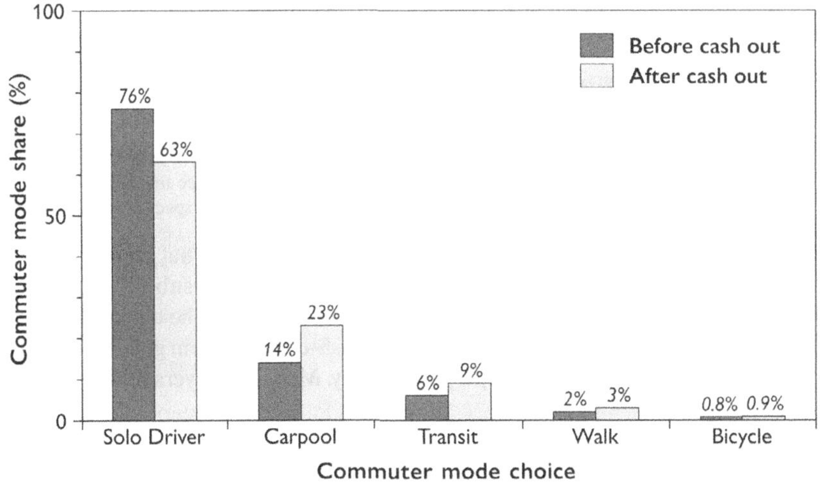

Fig. 7.4. Parking Cash-out Changes Commuters' Travel Choices. *Source*: Author.

work. Three times more solo drivers switched to carpools than to public transit, showing that cash-out works well even where public transit is not available. Parking cash-out takes advantage of empty seats in cars already on the road to work.

The shifts from solo driving to alternative modes reduced vehicle travel to work by 12%, equivalent to removing from the road one of every eight cars driven to and from work. Vehicle travel declined by 652 miles (1,050 km) per employee per year, and CO_2 emissions fell by 800 pounds (364 kg) per employee per year. A simple, fair and almost costless reform provided many benefits.

Parking cash-out is flexible. Some commuters might trade a parking subsidy for better health insurance. Some might put the parking subsidy into a retirement account, converting their free parking into higher retirement income. Other commuters might take the parking subsidy in cash and put it towards renting a more expensive apartment within walking or biking distance to work.

California's cash-out requirement is cheap for employers because it applies only to spaces rented from a third party. When a commuter cashes out free parking, money the employer previously paid for parking instead pays the commuter's cash allowance, and the firm breaks even. The case-study firms' total payments for parking spaces *and* cash rose by 3%, a small price to pay for treating all employees equally.

Employers can cash-out parking subsidies at no cost by redistributing the existing commute subsidies equally among all commuters, independent of their travel choices. This redistribution will not cost employers anything but will reduce vehicle travel, save petrol, and be fair to all commuters.

The employers' representatives said cash-out is simple, easy to administer, and almost automatic:

> 'It's very simple. It's not difficult at all'.
>
> 'The cash-out program is really simple. It is very easy to administer'.
>
> 'It's a good hiring incentive for us'.
>
> '[Cash out] is an excellent recruiting point because people count it as income'.
>
> 'Employees are grateful and thankful and motivated. That's a plus for the company'.
>
> '[Cash out] made employees happy. We emphasize it in our new employee orientation'.

Parking cash-out is a simple variation on a traditional fringe benefit. Giving commuters more choices about how to receive this benefit will reduce employer-paid parking's economic and environmental costs. Income tax revenue will also increase when commuters voluntarily choose taxable cash instead of a tax-exempt parking subsidy.

COVID-19 has upended the assumption that all employees will commute to work every day and that parking is offered on a monthly basis. For employees who occasionally work from home, some employers now offer parking cash-out on a daily basis. Commuters can park for free every day they drive to work, but they get the equivalent cash value of free parking every day they come to work without a car. Commuters can cash-out every day by not driving, and they can make different decisions on different days. Employees who don't commute to work won't pay or receive anything. A recent study published by the US Department of Transportation estimated that offering parking cash-out in Los Angeles on a monthly basis would reduce commuters' vehicle travel by 9%, and offering it on a daily basis would reduce vehicle travel by 17% (Abou-Zeid et al., 2023).

Daily payments are important for all parking permit systems, not just for parking cash-out. With a monthly permit, the marginal cost of parking on any day is zero. But if the monthly price is converted into a daily price, the marginal cost of parking is more than zero. For example, the University of California, Davis, has converted its monthly permit system to daily payments, with a simple way to pay by mobile phone and several choices of where to park. The prices for parking range from $5.10 a day in the most convenient lots to $2.00 a day in the remote lots. Every day, drivers decide whether to drive to campus and where to park. The Davis slogan is, 'You pay for the days you park, not for the days you don't'.

Unfortunately, parking cash-out has fared poorly in California's bureaucracy. The California Air Resources Board (CARB) is responsible for enforcing the cash-out law, but an audit found that CARB had not even notified employers about the law (Hill, 2002). Some employers voluntarily comply with the law if they hear about it, and others comply because they think it is sensible and fair, but most ignore it. If California is serious about decarbonising transportation, it should enforce the 30-year-old parking cash-out law.

Employers who fail to comply with the cash-out law discriminate against commuters who ride the bus to work. In Los Angeles, the median household income of Metro bus riders is $18,000 a year, and 92% of the bus riders are people of colour (Hymon, 2020). Enforcing California's cash-out law will increase transit ridership and reduce economic and racial inequality. Failing to enforce the law shows a serious lack of concern for transit riders and social justice.

Rather than expecting every state to enact and enforce cash-out laws, the best policy would be to reform the US Internal Revenue Code's income tax exemption for employer-paid parking. The federal government subsidises employer-paid parking by exempting the subsidy from employees' taxable income. The Internal Revenue Code subsidises driving to work.

Because removing a tax exemption is politically difficult, the federal government could instead stipulate that an employer-paid parking subsidy is exempt from income taxes only if the employer offers commuters the option to cash it out. The parking subsidy will become taxable income if an employer does not offer cash-out (Shoup, 2005b; Shoup & Pickrell, 2021). A bill requiring parking cash-out as a condition for the tax exemption of a parking subsidy, H.R. 8555, was introduced in Congress in 2022 and again in 2023.

By treating all commuters equally, parking cash-out reduces solo driving and increases transit ridership, carpooling, cycling, and walking to work. It also conserves petrol, improves air quality, and reduces carbon emissions. And employees' welfare increases without significantly increasing employers' costs. All these benefits result from giving commuters more choices about spending their own earnings.

Parking cash-out is a practical, fair, and almost costless reform. Subsidising people rather than parking will increase economic efficiency, environmental quality, and transportation equity.

CONCLUSION: YOU PAY FOR YOUR PARKING AND I'LL PAY FOR MINE

Parking reforms are the simplest way to start undoing a century of car-centred city planning. They are politically feasible – some cities have already adopted them – and studies of the results have found dramatic benefits with low or no costs.

All things considered, market rate parking is cheaper and fairer than free parking. Shifting the cost of parking to the parkers will make cities more expensive for cars and better for people.

NOTE

1. Editors' note: quote attributed to the University of Chicago economist Henry Simons; see Shoup (2011).

REFERENCES

Abou-Zeid, G., Grant, M., Heinrich, S., & Shah. D. (2023). *An assessment of the expected impacts of city-level parking cash-out and commuter benefits ordinances* [Report Number FHWA-HOP-23-023, 151 pages]. ops.fhwa.dot.gov/publications/fhwahop23023/index.htm

Brown, A., Mukhija, V., & Shoup, D. (2018, May 2–7). Garage apartments. *Zoning Practice*. https://anne-brown.com/wp-content/uploads/2018/11/brown_mukhija_shoup_2018_zoningpractice.pdf

Brown, A., Mukhija, V., & Shoup, D. (2020, March). Converting garages into housing. *Journal of Planning Education and Research*, *40*(1), 56–68.

Chrisafis, A. (2024, February 4). Parisians vote in favour of tripling parking costs for SUVs. *The Guardian*. https://www.theguardian.com/world/2024/feb/04/parisians-vote-in-favour-of-tripling-parking-costs-for-suvs#:~:text=Parisians%20have%20voted%20to%20triple,heavy%2C%20large%20and%20polluting%20cars

City of Seattle. (2023). *Accessory dwelling units 2022 annual report*. City of Seattle.

Court, A. (2022, October 26). I park my car wherever I want because I'm a millionaire – Only broke people care about tickets. *New York Post*. https://nypost.com/2022/10/26/i-park-my-car-wherever-i-want-because-im-a-millionaire/

Gray, N. (2024). *California ADU reform: A retrospective*. California Yimby Environment Fund. cayimby.org/wp-content/uploads/2024/01/CAY-ADU_Report-2024-v4.pdf

Hill, E. (2002). *A commuter's dilemma: Extra cash or free parking*. Legislative Analyst's Office. lao.ca.gov/Publications/Detail/882

Hymon, S. (2020). *Metro programs that work toward racial justice*. Los Angeles Metropolitan Transportation Authority.

Jacqout, G. (2022, March 11). L'obligation de pose de panneaux photovoltaïques sur les grands parkings extérieurs adoptée au Sénat. *Public Senat*. https://www.publicsenat.fr/actualites/environnement/l-obligation-de-pose-de-panneaux-photovoltaiques-sur-les-grands-parkings

Jonlin, A. (2023). Cutting the curb: Driveways and the right of access. *New York University Journal of Legislation and Public Policy*, *25*(3), 837–874.

Kassam, A. (2014, April 30). Madrid's smart parking meters to charge more for most polluting cars. *The Guardian*. https://www.theguardian.com/world/2014/apr/30/madrid-smart-parking-metres-polluting-cars

Millard-Ball, A., West, J., Rezaei, N., & Desai, G. (2022). What do residential lotteries show us about transportation choices? *Urban Studies*, *59*(2), 434–452.

Pinski, M. (2018). *Bundled parking and travel behavior*. https://escholarship.org/uc/item/56b0r9f3

INTRODUCTION

This chapter highlights the major issues facing the flexible transport sector (see Mulley & Nelson, 2016) and considers its potential contribution to the net zero agenda. Flexible transport services (FTSs) are defined here as collective services provided for passengers that are flexible in one or more of the dimensions of route, vehicle allocation, vehicle operator, type and time of payment, and passenger category. Each of the dimensions can vary along a continuum of demand responsiveness, from services which, on the one hand, are fixed some time before operation (the fixed route), to services where, on the other hand, the dimensions are determined closer to the time of operation (such services are often commonly referred to as demand responsive transport (DRT)). This chapter encompasses all aspects of flexible transport in contrast to what might be perceived as a narrower focus on paratransit, which in certain contexts is strongly associated with health and special needs. With the exception of the discussion related to informal transport in developing countries, here the term 'paratransit' is reserved for the FTSs in the United States.

For flexible transport, the world has changed dramatically in recent years with consequential changes to the external factors which affect operators. Shared transport, the central tenet of flexible transport, has taken a wider meaning. In 2025, shared transport is commonly understood in the context of car sharing where there is differentiation between commercially based shared car hired by the hour (such as 'Go Get' in Australia), community-based peer-to-peer car sharing (such as 'Car next door'), and carpooling, often used for commuting, where the owner of the car travels with pre-arranged passengers to an agreed destination. The shared economy has grown and now encompasses more than just four-wheeled vehicles. Shared bike schemes have proliferated with more recent additions including electric bikes, and micromobility more generally has appeared in recent years, and many cities have shared e-scooter programmes (as well as an upsurge in the use of personal e-scooters – see Chapter 4). The consequence of the growth of the sharing economy in this way means that there is a considerable blurring of the boundary between public and private asset ownership.

Health issues have greater prominence now than a decade ago. This can partly be attributed to the COVID-19 pandemic but also to a greater emphasis on the issues that concern public health agendas and a recognition of the role of transport in addressing these. Interestingly, flexible transport, with its ability to offer pre-booking, vehicle occupancy, cashless payment, and real-time updates, exhibited an early bounce-back following lockdown and offered an indication of what conventional public transport might offer in the future. Public health guidance encourages undertaking greater activity to reduce the burgeoning of the lifestyle diseases of diabetes and obesity, and shared bike schemes have a role here. Pop-up bike lanes and other changes to enhance the active travel that occurred during and after lockdown of the COVID-19 pandemic have reinforced this message.

Technology, also, has dramatically changed over the last decade. Ten years ago, flexible transport was benefiting from the development of technology to support a more interactive connection between users and suppliers. But current-day technology was not envisaged. The extensive use of smartphones, together with a proliferation of transport apps, means that it is easier than ever to connect users and suppliers. The growth in geospatial software has also enhanced the ability of the operator to plan and execute their services (including guaranteed connections) in real time – a significant change from the earliest 'dial-a-ride' services when booking, by telephone only, may well have been required one week in advance.

Alongside this, flexible transport has developed to service a wider range of consumers than ever before, in contrast to earlier concentration of services in rural and periurban contexts. Flexible transport is more regularly apparent in urban environments with access for all services – a far cry from the dedicated special needs programmes which were seen as extraordinary when flexible transport was first introduced. Some of these urban services replace fixed routes in areas of low demand, whereas others provide first and last mile access to more conventional public transport modes and so the application of flexible services has widened over time. To the extent FTSs replace or reduce private car use and deploy resources more efficiently in the delivery of public transport, there is the potential to contribute to the net zero agenda.

Against this background, this chapter looks at eight issues in more detail. The first is the transitory nature of many new FTS schemes, either as pilots or those which do not develop into ongoing services because they 'fail' and cease to operate. The second is the role of the volunteer within flexible transport schemes, and the third issue is that of the challenges for special needs services. Fourth is the contribution of technology and, in particular, what technology has added to the provision of FTSs, and the fifth issue relates to the way in which changes in the shared transport landscape have impacted on the provision and take up of flexible transport. Sixth is the role of marketing, since flexible transport schemes tend to not be overly visible, and seventh is the nature and scale of FTS schemes in developing countries. The final issue is the blurring of the characteristics of urban and rural schemes, especially in the context of MaaS offerings. This chapter concludes with a discussion of the potential contribution of FTSs to achieving net zero in the transport sector.

THE STABILITY OF THE FLEXIBLE TRANSPORT MARKET

While some FTS schemes are long-standing, suggesting a degree of stability is possible, many new ones are transitory in nature, either as pilot schemes which do not develop into ongoing services or schemes which 'fail' and cease to operate. The issue of 'failure' in the FTS sector was investigated by Currie and Fournier (2020), who took a rather narrow focus on costs and duration of life and concluded that 50% of schemes last less than seven years. One example of a high profile 'failure' is *Kutsuplus* in Helsinki which was operational from October 2012 to December 2015 (Jokinen et al., 2019; Nelson & Wright, 2021). The service started with 10 microbuses, with an additional five vehicles added in November 2013. Heralded as an example of a technology-led solution (incorporating automated vehicle location, trip combination optimization, vehicle routing and travel time estimates, and web and SMS bookings), the service was stop to stop (both pre-defined and virtual). The reported subsidy was (€16/trip) with the average fare being almost double that of public transport. Weckström et al. (2018) report that it ceased due to budgetary constraints and (significantly) through the lack of in-depth understanding of user perspectives.

Almost 20 years ago, Brake et al. (2006) identified three key factors to be aware of when setting up an FTS service, namely: the requirement to develop an economic framework (where *Kutsuplus* patently failed – see Jokinen et al., 2019); the need to implement appropriate service design which is ideally developed with key stakeholders (which must include the users, another weakness in the Helsinki example); and awareness raising (see further discussion of marketing below). A proper understanding of the need for a robust economic framework is particularly critical: if the service provider does not have a clear picture of the costs and revenues associated with a particular operation, as well as an appreciation of avoidable costs associated with different time horizons, then prospects for success will be more limited. *Callconnect* in the United Kingdom is an example of a well-designed successful FTS service which was established in 2001 and has operated on a continuing basis ever since (see https://lincsbus.info/callconnect/). Nelson and Wright (2021) reported that since inception, the service has resulted in the level of unmet need in the communities served being reduced by 90% with the subsidy required for DRT approximately the same as that required for the former (less effective) fixed-route services. Key to success has been the lower costs of operating smaller vehicles and an acceptance that services operating in deeply rural areas will require high subsidy per passenger.

THE ROLE OF VOLUNTEERS

FTS schemes have a long tradition of working with volunteers to deliver a quality service. The earliest dial-a-ride services date to the late 1960s in the United Kingdom and were set up with an exclusive remit to provide transport services, based on strict eligibility criteria, for those who would not be able to meet their mobility requirements under other means such as via conventional public transport. Most special needs schemes remain 'closed' – i.e., they are restricted to those with special needs – notwithstanding the potential for movement towards general use as recognised by Emele et al. (2016) (see further discussion in the next section). A number of metropolitan-based schemes have achieved considerable scale, and Westerlund's (2016) detailed description of the 30 largest schemes worldwide includes several that belong to the 'million trip' club (for example, the NYC MTA paratransit programme operated 6.9 million trips in 2022). At the same time, many volunteer-based special needs schemes operate in more rural communities. The strap line of the rural Bürgerbus transport services in Germany, for example, is 'citizens drive citizens' to provide local public transport services using small buses (Schiefelbusch, 2016). A critical success factor noted by Schiefelbusch is the need to understand volunteer motivation, and this remains pertinent in 2025 as recruiting and retaining volunteers is becoming increasingly difficult.

While such community transport (CT) schemes can encompass a wide variety of local, assisted, and specialised transport services, they generally sit at the nexus between health, aged care/social service, and transport. Crucially, it is more about community than passenger transport with a focus on person-centred transport solutions often delivered door to door by community-based volunteers. Operating models span community-led not-for-profit, private, and government-run services. For example, CT in New South Wales (NSW), Australia, delivered 1.59 m trips to 71,000 customers (FY 22/23) via 68 providers with a combined fleet size of 900+ vehicles at a contract value of AU$91.1m. Many older people rely on these services for health, well-being, and social connection and as a social service supporting independent and healthier lifestyles, and indeed Nelson and Caulfield (2022) suggest that the CT sector should be recognised for the 'lifeline' services they provide to rural communities. Hensher et al. (2024) have recently proposed a community sharing model, the Car Community Club (CCC), targeted at harnessing underutilised car capacity to meet short, and particularly long-distance, journeys. The purpose of the club is to match private car trips between drivers and potential passengers: in this case, the drivers are volunteers with their donation partially offsetting the club's costs.

THE FLEXIBLE TRANSPORT LANDSCAPE FOR SPECIAL NEEDS

According to the World Health Organization (2018), over 15% of the world adult population have some form of disability. However, the term 'special needs' covers a wider spectrum of the disadvantaged population than those with visible disabilities. Special needs include those with mobility impairment, visual impairment, and those with psychological or psychiatric profiles. Those individuals with special needs typically suffer from transport disadvantage and social exclusion with significant lack of mobility which in turn leads to a lowered quality of life (Jones et al., 2018; Lucas & Musso, 2014). In the last 10 years, there has been a greater recognition of the spectrum of transport difficulties experienced by the special needs population together with a widening of the understanding of the range of disabilities that constitute special needs (Kassens-Noor et al., 2021).

Many people with special needs rely on public transport for their mobility, and there remains a need to improve the accessibility and availability of services for these individuals. The situation of people with special needs who could feasibly use public transport is significantly worse in car-dominant cities, although even in those cities that are considered 'transit-orientated', such as London, Hong Kong, and Singapore, older people in particular form a much lower share of the ridership than might be expected. For example, in Hong Kong, where car ownership is under 15% and 93% commute by public transport, the trip rate of older adults was only 74% of that of their younger counterparts (Szeto et al., 2017; see Chapter 17).

Perhaps the greatest change to the provision of flexible transport for people with special needs has come in the domain of CT. In the United Kingdom, as

to deal with projected increases in demand. The ambition is for users to enjoy optimum co-modality by 2040, with good connections between the various networks via multimodal hubs (for passengers and freight), and close coordination of infrastructure and spatial development. An alternative is the relatively new phenomenon of TNCs which originally entered the transport landscape as a platform-based solution to offer a convenient taxi-like service with short wait times (Koffman, 2016). Since their arrival in San Francisco in 2010, TNCs have transformed the nature of the for-hire market with services provided by well-known brands such as Uber, Didi (China), and Ola (India). In 2014, there were no TNC trips in either New York City and Chicago, but TNCs surpassed taxi ridership within two to three years and by 2019 provided 77% of the combined taxi/TNC trips in New York and 87% of such trips in Chicago. In the same period, the total number of for-hire trips increased by 100% and 220% in New York and Chicago, respectively (Teal, 2025).

Koffman (2016) reflected on the likely impact of TNCs as a disruptor to the paratransit sector, citing issues of insurance coverage and oversight of vehicle condition, driver suitability, and training. As was anticipated Uber have entered the paratransit market partnering with some US transport agencies to deliver their paratransit services. There are obvious limitations since TNCs cannot accommodate all types of Americans with Disabilities Act (ADA) paratransit trips beyond curb-to-curb and are mostly limited to same-day service offerings, many of which are still in the pilot phase (Ashour & Shen, 2022). A further US study found that paratransit services face myriad barriers to competing with TNCs. Managers of these services did not begrudge TNCs success so much as they were frustrated by their own inability to evolve in tandem (Gleason & Silverstein, 2023). Finally, it is noted that Uber have also positioned themselves in the wider transport market with products such as Uber Access which make the use of Uber WAV (wheelchair accessible vehicles) and Uber Assist (targeted towards older persons).

THE ROLE OF MARKETING

FTSs tend to be less visible than regular public transport services as they typically do not traverse the same route on a regular basis. Awareness raising is therefore a critical step in the development and promotion of FTS schemes, and part of this process has to include stakeholders who are in close contact with potential end users such as political representatives, church, and community groups. Awareness raising or marketing should not rely on a single tool, and the evidence suggests posters, public meetings, media events and press releases, promotional leaflets, and branded vehicles all have a role to play. Although often word of mouth is the most powerful marketing tool, it is important in planning FTS schemes to ensure the marketing budget is sufficient to make them visible on an ongoing basis. It is the greatest false economy to have too small a marketing budget.

In Australia, NSW is unique in providing one of the largest contemporary FTS programmes. Championed by a Transport Minister who advocated technology-led solutions, 20 on-demand transport trials, 11 urban and 9 regional, were implemented.

In urban areas, the trials focused on locations where the fixed route was not performing well, or where off-peak services were sub-optimal, or where first and last mile connections could increase patronage on the core network. Not all the trials have continued but one zonal service, 'Cooee' – a stop to stop, all day weekday service providing access and first and last mile journey – has been particularly successful, in part due to its intensive awareness-raising programme. Cooee's operator provided customer education with the intention of making the customer feel 'special' and a dedicated operation centre that passengers could connect with and which could contact the passenger. The trial generated many more trips per hour than projected (11.5 over 4.5) and carried more passengers per hour than expected (over 80 compared to the projected average of 25) and has moved to being a permanent service. Marketing and awareness raising certainly played a part in its success.

Over the last 10 years, FTSs have had access to a wider set of awareness-raising tools due to advances in technology. The increase in smartphone penetration has meant that operators can market and inform passengers about their services using apps. Similarly, there have been improvements in dispatching of passenger requests and the routeing of vehicles so that passenger requests can be met in real time rather than be booked some time ahead of the desired service. The Cooee service, for example, provides booking at the time the passenger wants to travel with an average wait time of 12 minutes in 2019 (L.E.K., 2020). Still, while technological advances have increased the ability of operators to make aspects of their services better known to passengers, comprehensive marketing strategies remain essential. All too often they are undervalued and consequently under-resourced.

FLEXIBLE TRANSPORT IN DEVELOPING COUNTRIES

There are examples of flexible transport at scale across Latin America, Asia, and Africa. Informal transport services, also called 'semi-formal', 'paratransit', or 'popular' transport, are offered by private individuals or groups (Kustar et al., 2023). Examples include *jeepneys* in the Philippines, *matatus* in Kenya, *dala-dalas* in Tanzania, *tro-tros* in Ghana, *trufis* in Bolivia, and *tuk-tuks* across South and Southeast Asia. Bruun and Behrens (2016) provide an extensive description of paratransit in sub-Saharan Africa, noting that the jitneys there achieve some remarkable mode share of total public transport (e.g., Nairobi, 87%, and Dar es Salaam, 98%). Key characteristics of these services are that they are often not planned by a centralised government authority or are managed under only limited regulations. They are regularly characterised by unscheduled public transport and for-hire services operating in whole, or in part, within the informal economy. In many cases, this informal transport is a response to gaps left by formal public transport undertakings (Behrens et al., 2021).

A persuasive argument on the positive aspects of informal transport at scale is put forward by Finn (2012), who points out that in the Global South, this is often the primary form of public transport, carrying more people every day than extensive metro systems in Western cities. Finn (2012) suggests that the constructive aspects of informal paratransit are usually overlooked, including their organisational capacity, their ability to make self-financing business, and their ability to identify and respond to user needs as cities and travel patterns evolve.

While the informal paratransit sector in Africa, Asia, and Latin America is heterogeneous, services are usually operated by small businesses organised into associations that exert varying degrees of self-regulation (Behrens et al., 2021). Over time, some of these informal services have become formal, while others have been abandoned by regulation or by replacement by a new mode, such as bus rapid transit (BRT). For example, in South Africa, the introduction of BRT was used to reform the public transport sector: many informal transport services were simply replaced by BRT which ultimately led to a growing recognition of the need for a policy context that focuses on complementarity between scheduled services and unscheduled minibus operations, as well as innovations to improve, restructure, and engage, rather than simply replace, minibus services (Schalekamp & Klopp, 2018). Other informal services have consolidated their activity aided in part by the widespread availability of mobile communications and digital platforms which have the potential to significantly influence operations.

In countries where private vehicle ownership is low but growing, having convenient and affordable public transport can help stem rising transport emissions (Kustar et al., 2023). However, the informal sector is still characterised by older, polluting vehicles, and therefore, a significant future challenge includes the greening of the fleet through electrification.

MAAS

While definitions vary, MaaS has several dimensions. It is app based (so requiring a smartphone), providing a journey planner that in real time provides mobility service options tailored to the user's requirement using multimodal options, as well as providing for payment at the point of use or with pre-purchased mobility bundle and providing the ticket for travel. The concept of MaaS was to reduce the need for personal ownership of a vehicle by promoting mobility using shared assets. The vision of MaaS was therefore to reduce reliance particularly on the car, with consequential sustainability benefits such as reduction in emissions and congestion, although to date there are no MaaS offerings that quite meet the aspirations of their early promoters. Whim (from MaaS Global), originally based in Finland, was providing the most extensive offer, but in 2024, it filed for bankruptcy having run out of venture capital (Intelligent Transport, 2024).

The multimodal focus of MaaS utilises a diverse array of services – from conventional public transport, taxis, car sharing, TNC services (such as Uber, Didi, Ola) and taxis – to provide the ultimate in flexibility. MaaS has received much more attention in an urban setting with pilots testing different aspects of potential longer-term operation (see Hensher et al., 2020). However, flexible transport schemes are less present in the highly dense areas since there is sufficient demand to make fixed-route services a viable option. It is more likely that FTSs can contribute to MaaS offerings in suburban, regional, and rural areas, although rural pilot projects have been short-lived, with none so far becoming permanent (Mulley et al., 2023). For FTSs to make a real contribution to MaaS, there needs to be a coordinated approach to the multimodal offer, typically thought to use public transport as its 'backbone' with other modes providing 'first and last mile' services for the traveller. Given the technology now exists to book FTSs in real time, FTSs have an unrealised potential to provide mobility services within a MaaS framework, and undoubtedly, this is in part due to the lack of realised success for MaaS more generally.

CONCLUSION: THE POTENTIAL CONTRIBUTION OF FTSs TO THE NET ZERO AGENDA

Technology advances have made transport services smarter and more able to meet the varied needs of users in real time. However, the point of FTSs is to deliver mobility. Such mobility is more than simply providing trips – it allows social interactions and social inclusion – but in its consumption can have a negative impact on climate change, especially in the absence of electric vehicle fleets (see Chapter 5). It is principally in the context of FTSs substituting for private car journeys that the environmental harm of travel can be reduced but not eliminated. As with other forms of public transport, FTSs make a significant contribution to the inclusiveness of society in their delivery of accessibility to those citizens who are transport disadvantaged.

The literature clearly identifies significant barriers to the implementation and expansion of FTS schemes. Despite the passage of time, many of these still exist, particularly the silo approach to the funding of public transport that pertains in many jurisdictions and hinders the partnership working which is fundamental to FTS scheme development. Understanding the costs of operation is also key to viable services in the future. Mulley and Clifton (2016) identified a decision-based approach to planning and operation of FTSs with the 'golden rule' which identifies how avoidable costs must be covered in the short term if a service is to be viable. The importance of applying this rule still stands, but the understanding of costs has improved, if only because technology improvements have allowed systems to be more sophisticated in the allocation of costs to activities.

Flexible transport schemes have the potential to grow with demographic change (particularly with the older age group expanding), changes in travel attitudes and choices and the impact of development patterns both within and beyond cities. This has been illustrated by the recent disruption caused by the COVID-19 pandemic. The experience of public transport during the pandemic was characterised by dramatic falls in patronage and a slow build back, particularly in the case of bus patronage (Nelson et al., 2022). Flexible transport has proven itself over many years to be a convenient and user-friendly form of transport. This may in part explain the situation in NSW and elsewhere (e.g., South Australia) where FTSs recovered more quickly than conventional fixed-route public transport (Nelson & Caulfield, 2022). Also relevant in other locations were other factors

Szeto, W. Y., Yang, L., Wong, R. C. P., Li, Y. C., & Wong, S. C. (2017). Spatio-temporal travel characteristics of the elderly in an ageing society. *Travel Behaviour and Society*, *9*, 10–20.

Teal, R. F. (2016). Technology changing market feasibility. In C. Mulley & J. D. Nelson (Eds.), *Paratransit: Shaping the flexible transport future* (pp. 333–356). Emerald Group Publishing Limited.

Teal, R. F. (2025). Transportation Network Companies. In C. Mulley, J. D. Nelson, & S. Ison (Eds.), *Elgar encyclopedia of transport and society*. Elgar.

Weckström, C., Ullah, W., Mladenović, M. N., Nelson, J. D., Givoni, M., & Bussman, S. (2018). User perspectives on emerging mobility services: Ex post analysis of Kutsuplus pilot. *Research in Transportation Business and Management*, *27*, 84–97.

Westerlund, Y. (2016). Development and status for large-scale demand responsive transport. In C. Mulley & J. D. Nelson (Eds.), *Paratransit: Shaping the flexible transport future* (pp. 53–74). Emerald Group Publishing Limited.

World Health Organization. (2018). *Disability and health*. https://www.who.int/news-room/fact-sheets/detail/disability-and-health

CHAPTER 9

TOWARDS NET ZERO RAILWAYS FOR NET ZERO TRANSPORT

John Preston and Simon Blainey

University of Southampton, UK

ABSTRACT

As a low-friction transportation system, rail has some strong environmental credentials. However, it also has a reliance on some carbon-intensive technologies and materials that make a transition to net zero problematic. This chapter examines the prospects for net zero railways through the lens of Holden et al.'s (2020) nine grand narratives for sustainable mobility. These nine narratives are the product of three stakeholder groups (specified here as the rail industry, government bodies, and end-users) and three intervention types (improve, shift, and avoid). Overall, it is argued that a mix of measures will be required to deliver net zero railways (and in turn net zero transport) that also meet wider sustainability goals, and that this mix will be context dependent.

Keywords: Electrification; end-users; governance; rail; whole life costs

INTRODUCTION

The concept of net zero is a product of the early 21st century, while the railways as we know them were a product of the 19th century. The challenge for what is for the most part a legacy system is to re-engineer railways, to adjust their operations, and to exploit their inherent advantages in terms of net zero. Steel wheel on steel rail is a low-friction system that has economic advantages over other modes

Towards Transport Net Zero
Transport and Sustainability, Volume 20, 129–141

ISSN: 2044-9941/doi:10.1108/S2044-994120250000020009

of transport in terms of carrying heavy loads and dense traffic at high speeds. It allows an energy efficiency per traffic unit that other modes cannot match, with commensurate environmental benefits, although as Smith (2003) noted this environmental advantage has reduced as rival modes have innovated at a greater pace than rail.

Moreover, railway assets may be long lived (for earthworks, design lives of 120 years are common and practical lives even longer), and there has been a tendency to use carbon-intensive materials (steel, concrete) and methods for their construction and maintenance. As a result, rail systems may have relatively high levels of embedded carbon (see, for example, Milford & Allwood, 2010), which may offset the relatively low emissions during operations (Chester et al., 2010; Pritchard & Preston, 2016). The challenge then becomes one of translating principles into practice, in terms of designing interventions that minimise the whole life-cycle carbon costs of infrastructure and rolling stock, while maximising the benefits to end-users.

A net zero railway would help meet the crisis of the climate emergency both directly by reducing rail's carbon footprint to zero and indirectly by permitting large-scale modal shift (which will require major increases in capacity) and by promoting denser and more energy efficient land use in urban areas (Cervero, 1998; Hall & Hass-Klau, 1985). However, it is not clear how such a railway will be delivered. A possible framework is provided by an adaptation of the nine grand narratives of Holden et al. (2020). These nine narratives are the product of three stakeholder groups (specified here as the rail industry, government bodies, and end-users) and three intervention types (improve, shift, and avoid). This is illustrated in Table 9.1, and the contribution of the three stakeholder groups is discussed in the next three sections. However, these divisions are not clear-cut, and the boundaries in practice are more fuzzy than those depicted in Table 9.1. For instance, land use is not simply the domain of the government, but of private developers, industry (including rail itself) and end-users. The boundaries between the government and the rail industry are not clear-cut in many cases.

THE RAIL INDUSTRY

For the rail industry, improvements include the use of renewable energy technologies for traction (electricity, hydrogen, biofuels), the use of lightweight and

Table 9.1. Some Grand Narratives of Net Zero Railways.

	Improve	Shift	Avoid
Rail industry	Cleaner rolling stock and infrastructure	Abstraction of traffic	Peak flattening
Government	Organisation and regulations	Grants and subsidies	Compact and diverse land use
End-users	Special interest groups	Flight-shaming and night trains	Teleworking and teleconferencing

Source: Holden et al. (2020).

low-carbon materials in vehicle manufacture, digitisation to improve operational efficiency, life-cycle analysis to reduce embedded carbon in infrastructure, the adoption of circular economy principles, and careful use of offsetting. Shift interventions include service expansion (classic heavy rail, light rail, high-speed rail), fares and ticketing initiatives, and improvements to the customer experience to attract passengers from more carbon-intensive modes. Avoid strategies might include a reduction in the traditional peaks and a move towards a more flat-graph operation to maximise train occupancy levels. These three sets of measures are considered in turn.

Improve: Cleaner Rolling Stock and Infrastructure

As a low-friction, steel-on-steel transportation system, rail's energy efficiency, and environmental advantages are relatively well known (Smith, 2003). Some useful recent data for Great Britain are provided by Office of Rail and Road (ORR, 2023a). For the financial year 2022–2023, for passenger rail services, mean emissions were 33 g CO_2e per passenger km and 616 g CO_2e per vehicle km. However, this masks large variations across rolling stock types and train operating companies. Emissions from electric rolling stock varied from 134 to 528 g CO_2e per vehicle km, while emissions from diesel rolling stock varied from 1,104 to 1,729 g CO_2e per vehicle km. Aside from the disruption caused by COVID-19, emissions per passenger km have reduced over time, due to the reduced carbon intensity of electricity generation and, in the main, the greater energy efficiency of new rolling stock. These findings are consistent with earlier studies. For example, the Department for Transport (DfT, 2007) reported rail emissions that varied from below 50 g CO_2e per passenger km (for an electric Class 91 IC225) to over 100 g (for a diesel Class 221) (see also Esters & Marinov, 2014). ORR (2023a) also gives data for freight services. These indicate average emissions of 26 g CO_2e per net tonne km (varying between freight operating companies by between 21 and 31 g) and 564 g CO_2e per vehicle km (varying between freight operating companies by between 135 and 230 g for electric traction and between 524 and 967 g for diesel traction). Emissions per net tonne km have remained relatively constant over time, representing a reliance on diesel traction and relatively limited innovation. DfT (2023) provides some modal comparisons. For a passenger journey from London to Glasgow, the emissions from aircraft are estimated at 176 kg CO_2e; for an average passenger car, it is 89 kg, while for rail, it is 30 kg; and for coach, it is 21 kg. Again, this is consistent with earlier data (DfT, 2007) but with some indication that rail's environmental advantage is being eroded. The emissions per passenger for an average electric car between London and Glasgow are estimated at 29 kg.

The carbon emissions from rail are low, but they are not zero. However, given the advantages of electric rolling stock, the path to net zero would seem straightforward. The national rail network of Great Britain has a relatively low percentage of electrified lines (38% of route kms). Line electrification would seem the obvious solution, along with decarbonisation of electricity generation. For example, the *Traction Decarbonisation Network Strategy* (TNDS) advocated increasing the percentage of electrified lines to 90% (Network Rail, 2020). However, the capital costs are high, and delivery would be problematic. An alternative has been put forward by the Railway Industry Association (RIA, 2024), in which the percentage of route kms electrified is increased to 66% through the development of a rolling programme of electrification. This would be complemented by the replacement of approximately 1,100 sub-100 mph diesel multiple units (DMUs) with battery electric multiple units by 2030 and then the development of a rolling programme of rolling stock renewal. Unit costs would be reduced by ensuring a steady workflow and avoiding past booms and busts in electrification as a result of stop–start investment policies. Other options that were considered but dismissed include hydrogen electric multiple units (technology not yet proven although trials with Hydroflex are underway and green hydrogen availability is uncertain) and alternative drop-in fuels such as biofuels and synthetic fuels (likely to be prioritised for other sectors such as aviation but could be used for rail freight). Another possibility is intermittent or discontinuous electrification in which overhead line equipment (OLE) is not provided where there is insufficient clearance to do so (as a result of tunnels or road bridges, for example). However, this would require on-board energy storage and would increase the possible points of failure as the pantograph connects and disconnects with the OLE.

The discussion above has focused on operational carbon emissions (opex), which are related to the energy used to operate rail services. This is based on a well-to-wheel approach in that it takes into account the carbon used to extract, refine, and distribute diesel and the carbon emitted in generating and transmitting electricity. However, it does not take into account the embedded carbon in the production, use and disposal of rail infrastructure, and rolling stock (capex). An early study by Chester and Horvath (2009) suggested that such capex increased urban rail life-cycle energy consumption by 155%, although this was based on estimates for the Bay Area Rapid Transit (BART) system in and around San Francisco. Equivalent multipliers for the HS2 high-speed rail line in the United Kingdom have been estimated at 25%–35% (Network Rail, 2009).

Given this, there has been extensive research on reducing the embedded carbon inherent in rail systems. This involves less use of carbon-intensive materials such as concrete and greater use of recycling, for example, the use of double-headed rails (Milford & Allwood, 2010). This will involve moving from a linear model of production to a circular model (International Union of Railways [UIC], 2021). Research at the University of Southampton has considered the choice of materials for sleepers (Rempelos, Preston, et al., 2020), adjustments to ballasted track systems including under sleeper pads and random fibre reinforcements (Rempelos, Ortega Hortelano, et al., 2020), the use of stainless steel-reinforced concrete bar in structures (Moodley et al., 2022) and earthworks maintenance strategies (Armstrong et al., 2024; Najafpour Navaei et al., 2024). In isolation, these measures lead to only small improvements, although in combination they are more substantial. However, if net zero is to be achieved, new methods (and energy sources) are going to be required to get labour, equipment, and materials to site and to power the on-site equipment.

Decarbonising Transport: A Better, Greener Britain (DfT, 2021) outlined aspirations to deliver a net zero railway by 2050 (although there would be some emissions that would need offsetting – the report only estimated a 97% reduction in gross emissions by 2050) and the removal of diesel traction by 2040. However, detailed plans were not provided, and there has been some subsequent de-scoping of rail investments since then, not least the scrapping of HS2 north of Birmingham.

Such plans can only be achieved with the support of the rail industry. For example, the major rolling stock and rail equipment supplier Hitachi has a decarbonisation strategy which aims to achieve net zero of the company's own emissions (Scopes 1 and 2) by 2030 and net zero for its supply chain (Scope 3) by 2050. However, it should be noted that Scope 3 emissions are up to 90% of total emissions.[1] Other original equipment manufacturers (OEMs) (such as Alstom and Siemens) have similar plans. The plans for the rail infrastructure authority in Great Britain, Network Rail, are illustrated in Fig. 9.1. Such plans may be part of environmental management strategies; environment, social, and governance (ESG) frameworks; and corporate social responsibility (CSR). A recent example is the United Kingdom's Rail Delivery Group Green Travel Pledge that has developed a carbon calculator for 40,000 business travel routes. This indicates that on average carbon emissions are 94% lower by train than petrol car and is designed to influence corporate travel pledges.[2]

One other area to consider is the scope for digital technology to replace or enhance physical infrastructure. A textbook example is that of the European Train Control System (ETCS), where simulation work suggested that a move from unsupervised Level 1 to Level 3 (in essence a move from traditional signalling to in-cab signalling) could lead to a theoretical increase in capacity of 42% (UIC, 2008). However, there are at least two issues with this calculation. The first is that in practice Level 1 would be implemented with some limited supervision. In such instances, the capacity increase reduces to 28%. The second is that the

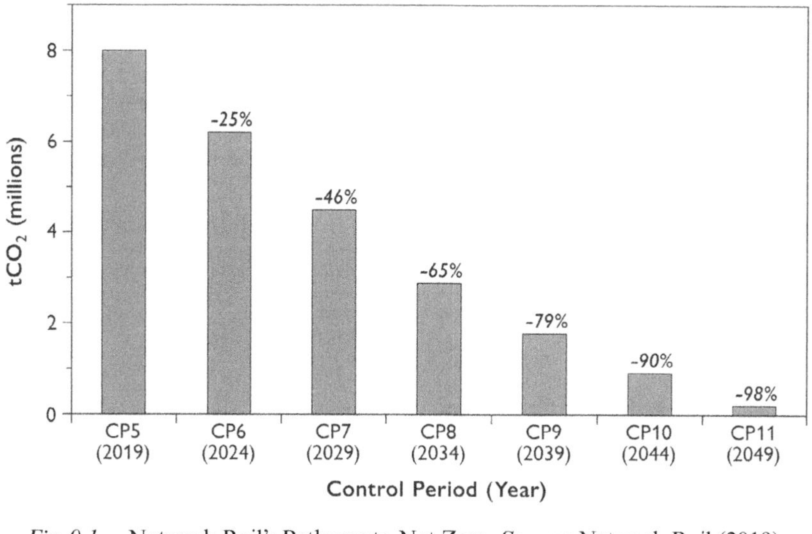

Fig. 9.1. Network Rail's Pathway to Net Zero. *Source*: Network Rail (2019).

calculations are for plain track. However, for a transport network, the capacity constraints are provided by the nodes (stations and junctions) not links. For many rail systems, the overall system capacity is determined by the turnaround times at the dead-end terminals. In practice, the capacity increases promised by the digital railway have proved elusive. Other work has shown how using digital tools to produce smart timetables and rescheduling can increase the resilience of timetables to disruption and obviate the need to invest in expensive (and carbon-intensive) infrastructure, such as grade-separated junctions (flyovers) (Chou, 2008).

Shift: Abstraction of Traffic

We have seen above that rail has environmental advantages over other modes of transport, and due to the nature of its infrastructure, it can electrify relatively quickly. Modal shift from other modes to rail should therefore help in the path to net zero. One way of delivering such modal shift is through the provision of new services. Table 9.2 shows that high-speed rail (HSR) can deliver a substantial mode shift as between 30% and 70% of HSR demand is abstracted from air and road, with the mean figure for the five schemes being around 48%.

For classic rail in the United Kingdom, the Institute for Transport Studies (ITS, 2016) estimated that between 21% and 35% of new demand was abstracted from car, depending on market segment. Similar abstraction rates pertain to light rapid transit. For a predominantly street-based system, such as the Sheffield Supertram, some 20% of demand was estimated to come from car. For the first line of the Manchester Metrolink, which was a replacement for a heavy rail service, the proportion of demand coming from cars was around 27% (Balcombe et al., 2004, Tables 12.11 and 12.15).

Another way of investigating the extent of modal shift is through the use of cross-elasticities. For example, early versions of the UK *National Transport Model* assumed an elasticity of car demand with respect to rail fares of 0.025. If rail fares reduced by 10%, car demand would fall by 0.25%. That seems miniscule but rail's share of trip kms is only around 10% of all travel, whereas for cars, it is around 80%. This shift from car to rail would lead to a 2% increase in rail use. Wardman et al. (2018) undertook a meta-analysis of 1,096 cross-elasticities from 93 studies. They found that cross-elasticities to time-based attributes tended to

Table 9.2. Abstraction Factors Resulting from the Introduction of HSR.

Route	Paris-Lyon[1]	Madrid-Seville[2]	Madrid-Barcalona[3]	Thalys[4]	Eurostar[4]
% HST traffic generated from:	1980–1985	1991–1996 forecast	'Before HSR' to 'After HSR'	Range not given	Range not given
Induced	29	50	20	11	20
Road	11	6	10	34	19
Conventional rail	40[a]	20	10	47	12
Air	20	24	60	8	49

Sources: [1]Bonnafous (1987); [2]de Rus and Inglada (1997); [3]Coto-Millán et al. (2007); and [4]Segal (2006).

[a]All Paris-Lyon's 'after' rail travel is presumed to be by HST (i.e., no conventional rail following introduction of HST), since alternative journey time is ~5 hours compared to ~2 hours by HST.

exceed those for price-based attributes, but cross-elasticities for car demand were low. This indicates that there is a limit to the extent of modal shift to rail, a finding that is echoed by more qualitative studies (Blainey et al., 2012; Stanton et al., 2013), at least within the constraints of no road user charging. Such abstraction that can be achieved might be facilitated by pricing policies that explicitly play on rail's environmental advantages, such as eco-passes in Germany and Switzerland and Klima tickets in Austria.

Abstraction rates and cross-elasticities are less common in rail freight. Instead, growth targets are more commonplace. For example, in December 2023, the UK government set a target to grow freight carried by rail by at least 75% by 2050. It was noted that a tonne of freight moved by rail produces around a quarter of the carbon emissions that it would if it were moved by road. One train could replace up to 129 heavy goods vehicles (HGVs).[3] This will be considered in more detail below under the 'Governments' heading, but such targets are also widely used by rail freight firms.

Avoid: Peak Flattening

Many passenger railways have a high dependency on commuting, which may account for as much as 50% of demand and be concentrated in a four-hour period, out of an operating day that can be around 18 hours. The result is high levels of crowding in the peak and high utilisation of infrastructure which can lead to problems with punctuality and reliability. For example, modelling work at a busy station indicated one additional train (representing a 1.8% increase in service) would lead to an increase in delays of 4.7%, an elasticity of delays to service levels of around 2.6 (Preston, 2017). Such heavily peaked operations can perform poorly in economic terms but also have adverse consequences in terms of the environment, with poor rolling stock and infrastructure utilisation over the operating day. A flat-graph operation (in which the same number of trains per hour is provided throughout the operating day) may have economic and environmental benefits. As we will detail later in this chapter, the growth in remote working and reduced demand for commuting and business travel may provide opportunities to flatten the peak over and above the use of pricing premiums in the peaks (and discounts in the off peak).

GOVERNMENTS

For governments, there is scope for improvements in the way they interact with the rail industry in terms of more appropriate organisational and regulatory structures. Modal shift may be achieved by appropriate levels of funding and finance, investments in first and last mile access, and multi-modal integration. Governments may also promote compact and diverse land use and flexible working practices that reduce the need for travel.

Improve: Organisation and Regulation

Where railways are under public ownership and control, as is still true for much of the world's railways including China and India, ensuring railways achieve net zero can, in theory, be delivered through a system of governmental command and control. However, this public ownership model has failed in many places, not least in sub-Saharan Africa and Latin America (Resor & Laird, 2013), and has been replaced by concession agreements. Net zero provisions could be written into such concessions, and hence, governance by contract is feasible. Where markets are liberalised, ensuring the achievement of environmental objectives may be secondary to commercial objectives. Targets could, however, be set by an independent regulator, which could be industry based such as the ORR in the United Kingdom or could be cross-sector such as the Environmental Protection Agency in the United States. In such cases, governance by steering rather than by rowing can be adopted (Rhodes, 1996) or in rail parlance through a guiding mind rather than a directing mind.

Shift: Funding and Finance

In many instances, governments fund railway activity. Implicit in this funding is the desire to keep traffic on the railways and, where possible, grow such traffic. A case in point is provided by the national railways in Great Britain. For the financial year 2022/2023, £11.9 billion of operational income came from the government, some 52% of total income (ORR, 2023b). In addition, the government provided some £9.2 billion of capital investments, including £6.9 billion for HS2, £2 billion for Network Rail enhancements (including the Transpennine Rail Upgrade, East Coast Digital, and the Midland Mainline Upgrade), and £0.1 billion for East-West rail. Further funding may be provided for feeder services to/from stations, such as electric buses, to solve the first/last mile problem. What is noticeable is how little of this funding is specifically tied to environmental objectives. An exception (albeit on a small scale) is the modal shift revenue grant scheme (MSRS) that is designed to support modal shift from road freight to rail (and water). It has a budget of around £20 million a year and in 2022–2023, helped remove 900,000 lorry journeys from Britain's roads, saving almost 40,000 tonnes of carbon dioxide.

Avoid: Compact and Diverse Land Use

Perhaps the area where governments have both the greatest opportunity to influence the decarbonisation of travel and have had the most negative impact in practice is through land-use planning. A sustained and universal policy of increasing land-use densities, prioritising mixed-use developments and focusing development around public transport hubs could both reduce the need for travel and encourage remaining trips to be made by public and active travel. Such policies are exemplified by the '15-minute city' movement as advocated for by Carlos Moreno (Moreno et al., 2021) and progressively implemented in some urban areas, notably Paris under the leadership of Anne Hidalgo. This aims to deliver neighbourhoods where residents can access all essential services within a 15-minute active travel journey of their home. Ideally, such neighbourhoods should be centred on a railway station or other public transport hub so that longer trips could be made by rail, removing the need for car ownership. While positive in terms of environmental sustainability, 15-minute city policies have been criticised for leading to gentrification if not applied universally across a city and have also been the target of online conspiracy theories. The integration of such land-use patterns in existing urban areas can be challenging, although the Urban Transport Group (2022) has suggested the use of 'gentle densification' around existing transport hubs as a first step. More generally, in order to minimise embedded emissions from the construction of transport infrastructure, governments should maximise the use of existing infrastructure as hubs for new development by basing new settlements around disused (but intact) railway formations or underused stations or lines on the existing rail network.

Governments also have the ability to both promote and mandate (via taxation and/or regulation) virtual alternatives to travel, such as home and flexible working and online access to retail and other services. Such policies were implemented effectively in many countries for public health reasons during the COVID-19 pandemic. However, the long-term sustainability gains from these policies have been compromised first by associated warnings to avoid public transport in favour of private transport for health reasons and subsequently (at least in the United Kingdom) by government campaigns aiming to get people to return to the office because of the perceived benefits for both productivity and urban economies.

Ultimately, government intervention to help achieve net zero could include the introduction of carbon taxes or personal carbon budgets which would both limit overall travel and encourage the use of lower carbon modes such as rail. While such policies might be effective in reducing emissions, it is not clear if or when they would ever be viewed as being politically acceptable in democratic countries, and they would almost certainly generate vocal opposition from 'populist' politicians. However, there is some evidence to suggest that the use of deliberative democracy techniques such as citizens' parliaments may be able to help provide a pathway for the implementation of more ambitious emission reduction policies (Willis et al., 2022).

END-USERS

While end-users may have limited direct power to change the services offered to them by the railway industry or to affect the sustainability of these services, through their behaviour, they have substantial potential to influence how services change over time, particularly if they can organise to have a collective impact.

Improve: Special Interest Groups

Both passenger and freight users of the rail network can champion service improvements via collective action, through groups for passenger services and Rail Freight User Groups for freight. For example, in Great Britain, community rail groups under the umbrella of the Community Rail Network (formerly ACoRP[4]) have undertaken some advocacy around increasing the sustainability of existing rail operations (particularly through electrification). There is also in theory the potential for the customers of rail freight operators to request the use

NOTES

1. https://www.hitachirail.com/csr-and-sustainability/decarbonisation-strategy/ (accessed 30 July 2024).
2. https://www.raildeliverygroup.com/uk-rail-industry/green-travel-pledge.htm (accessed 1 August 2024).
3. https://www.gov.uk/government/publications/rail-freight-growth-target/rail-freight-growth-target (accessed 31 July 2024).
4. Association of Community Rail Partnerships.
5. See, for example, https://www.facebook.com/@greatscenicrailways/?locale=en_GB

REFERENCES

Armstrong, J., Helm, P., Preston, J., & Loveridge, F. (2024). Economics of geotechnical asset deterioration, maintenance and renewal. *Transportation Geotechnics*, *45*, 101185.

Balcombe, R., Mackett, R., Paulley, N., Preston, J., Shires, J., Titheridge, H., Wardman, M., & White, P. (2004). *The demand for public transport: A practical guide* [Report 593]. TRL, Crowthorne.

Blainey, S. (2022, July 18–22). *Sleepers awake? The impact of market liberalisation on the evolution of overnight rail services in Europe*. Paper presented at the *International Geographical Union Centennial Congress*, Paris.

Blainey, S., Hickford, A. & Preston, J. (2012). Barriers to passenger rail use: A review of the evidence. *Transport Reviews*, *32*(6), 675–696.

Blainey, S., & Preston, J. (Eds.). (2022). *Sustainable railway engineering and operations*. Emerald Publishing.

Bonnafous, A. (1987). The regional impact of the TGV. *Transportation*, *14*, 127–137.

Cervero, R. (1998). *The transit metropolis. a global inquiry*. Island Press.

Chester, M. V., & Horvath, A. P. (2009). Environmental assessment of passenger transportation should include infrastructure and supply chains. *Environmental Research Letters*, *4*, 1–8.

Chester, M. V., Horvath, A., & Madanat, S. (2010). Comparison of life-cycle energy and emissions footprints of passenger transportation in metropolitan regions. *Atmospheric Environment*, *44*(8), 1071–1079.

Chou, Y.-H. (2008). *Distributed approach for rescheduling railway traffic in the event of disturbance* [PhD thesis, University of Birmingham].

Coto-Millán, P., Inglada, V., & Rey, B. (2007). Effects of network economies in high-speed rail: The Spanish case. *Annals of Regional Science*, *41*, 911–925.

Department for Transport. (2007, July). *Delivering a sustainable railway: A strategic direction for the rail industry*. DfT.

Department for Transport. (2021, July). *Decarbonising transport: A better, greener Britain*. DfT.

Department for Transport. (2023, October). *Journey emission comparisons: Interactive dashboard*. Department for Transport. https://maps.dft.gov.uk/journey-emission-comparisons-interactive-dashboard/index.html

de Rus, G., & Inglada, V. (1997). Cost-benefit analysis of the high-speed train in Spain. *Annals of Regional Science*, *31*, 175–188.

Esters, T. & Marinov, M. (2014). An analysis of the methods used to calculate the emissions of rolling stock in the UK. *Transportation Research Part D*, *33*, 1–16.

Hall, P., & Hass-Klau, C. (1985). *Can rail save the city? Impacts of rail rapid transit and pedestrianisation on British and German cities*. Ashgate.

Holden, E., Banister, D., Gossling, S., Gilpin, G., & Linnerud, K. (2020). Grand narratives for sustainable mobility: A conceptual review. *Energy Research & Social Science*, *65*, 101454.

Institute for Transport Studies. (2016). *Fares conditional elasticity study* [Final report]. Passenger Demand Forecasting Council (PDFC), London. Results available in WebTAG Data Book, Table A5.4.5.

International Union of Railways. (2008). *Influence of ETCS on line capacity – Generic study*. UIC.

International Union of Railways. (2021). *UIC sustainability: Circular practices in the railway and ways forward: Reuse project final report*. UIC.

Milford, R. L., & Allwood, J. M. (2010). Assessing the CO_2 impact of current and future rail track in the UK. *Transportation Research Part D*, *2*, 61–72.

Moodley, H., Blainey, S., Preston, J., & Afshan, S. (2022, June 6–10). *Can stainless steel reinforcement bar facilitate more sustainable railway bridges? A comparative life cycle cost and carbon analysis*. Paper presented at World Congress on railway research 2022, ICC, Birmingham, United Kingdom.

Moreno, C., Allam, Z., Chabaud, D., Gall, C., & Pratlong, F. (2021). Introducing the "15-minute city": Sustainability, resilience and place identity in future post-pandemic cities. *Smart Cities*, *4*(1), 93–111.

Najafpour Navaei, T. A., Blainey, S., Powrie, W., & Preston J. (2024). Carbon footprinting of railway infrastructure: A standardised, consistent data collection method. *Carbon Management*, *15*(1), 2369939.

Network Rail. (2009, August). *New lines programme. Strategic business case*. Network Rail.

Network Rail. (2019). *Our ambition for a low-emission railway 2020–2050*. Retrieved February 9, 2021, from https://www.networkrail.co.uk/sustainability/a-low-emission-railway/

Network Rail. (2020, September). *Traction decarbonisation network strategy. Interim programme business case*. Network Rail.

Office of Rail and Road (ORR). (2023a, November). *Rail emissions, March 2022 to April 2023*. ORR. https://dataportal.orr.gov.uk/statistics/infrastructure-and-emissions/rail-emissions/

Office of Rail and Road (ORR). (2023b, November). *Rail industry finance. April 2022 to March 2023*. ORR.

Preston, J. (Ed.). (2017). *Developing integrated tools to optimise (DITTO) rail systems: Project deliverable 4.4. milestone 13* [Final report]. Rail Safety and Standards Board (RSSB), London.

Pritchard, J. A., & Preston, J. M. (2016). Sustainable railways: Trade-offs between operational and embedded energy. *Proceedings of the Institution of Civil Engineers – Transport*, *169*(5), 298–307.

Railway Industry Association. (2024). *Delivering a low cost, higher performing, net zero railway by 2050*. RIA. Retrieved November 23, 2024, from https://www.riagb.org.uk/ALCHPNZRW24

Rempelos, G., Ortega Hortelano, A., Blainey, S., Preston, J., Le Pen, L., & Armstrong, J. (2020). A method for assessing the life cycle costs of modifications to ballasted track systems. *Construction and Building Materials*, *263*, 120603.

Rempelos, G., Preston, J., & Blainey, S. (2020). A carbon analysis of railway sleepers in the United Kingdom. *Transportation Research Part D*, *81*, 102285.

Resor, R. R., & Laird, P. G. (Eds.). (2013). Railroad privatization and deregulation: Lessons from three decades of experience worldwide. *Research in Transportation Business & Management*, *6*(Special Issue), 1–122.

Rhodes, R. A. W. (1996). The new governance: Governing without government. *Political Studies*, *44*(4), 652–667.

Segal, J. (2006, September). *High speed rail – The competitive environment*. MVA Consultancy – Presentation given at European transport conference.

Smith, R. A. (2003). Railways: How they may contribute to a sustainable future. *Proceedings of the Institution of Mechanical Engineers, Part F: Journal of Rail and Rapid Transit*, *217*(4), 243–248.

Stanton, N. A., McIlroy, R., Harvey, C., Blainey, S. P., Hickford, A. J., Preston, J. M., & Ryan, B. (2013). Following the cognitive work analysis train of thought: Exploring the constraints acting on modal shift to rail transport. *Ergonomics*, *56*(3), 522–540.

Townsend, J. (2022). Community rail: Helping our railways to deliver social, environmental and economic sustainability (Chapter 4). In S. Blainey & J. Preston (Eds.), *Sustainable railway engineering and operations*. Emerald Publishing.

Urban Transport Group. (2022). *The good life: The role of transport in shaping a new and sustainable era for suburbs*. Urban Transport Group.

Wardman, M., Toner, J., Fearnley, N., Flügel, S., & Killi, M. (2018). Review and meta-analysis of inter-modal cross-elasticity evidence. *Transportation Research A*, *118*, 662–681.

Willis, R., Curato, N., & Smith, G. (2022). Deliberative democracy and the climate crisis. *WIREs Climate Change*, *13*, e759.

CHAPTER 10

FLYING TOWARDS NET ZERO: DECARBONISING AVIATION AMID A CLIMATE CRISIS

Evangelia Pantelaki[a] and Lucy Budd[b,c]

[a] *Emirates Aviation University, UAE*
[b] *De Montfort University, UK*
[c] *Institute of Transport and Logistics Studies, The University of Sydney, Australia*

ABSTRACT

In 2021 the international air transport industry committed itself to achieving net zero carbon emissions by 2050 in line with the Paris Agreement to keep global temperature rises below 1.5°C above pre-industrial levels. Yet, in 2022, as demand for air travel rebounded following the COVID-19 pandemic, air transport's CO_2 emissions reached almost 800 million tonnes, approximately 80% of the pre-pandemic level. Reconciling growing consumer demand for flight – a mobility regime which still relies on finite fossil fuel as its primary energy source – with net zero commitments in the midst of a climate crisis is not only a key challenge for the air transport sector but for global society as a whole. This chapter aims to review the key technological propositions (including sustainable aviation fuels (SAFs), hydrogen, and electric/hybrid aircraft) and international regulatory interventions which aim to facilitate commercial aviation's transition towards net zero. This chapter concludes by assessing the key challenges to meeting aviation's 2050 net zero ambition.

Keywords: Climate change; net zero: air transport; sustainable aviation; decarbonisation; airport operations

Towards Transport Net Zero
Transport and Sustainability, Volume 20, 143–157

ISSN: 2044-9941/doi:10.1108/S2044-994120250000020010

actions of third-party contractors and suppliers (which include, but are not limited to, in-flight catering providers, airport hotels, and retailers). However, as later sections of this chapter show, defining 'ownership' of emissions and delivering multi-stakeholder agreements on how to drive, audit, and deliver the necessary carbon reductions is complicated.

Although there is growing scientific and political consensus on the need to urgently reduce aviation's GHG emissions, there is no consensus on how best to achieve it. Furthermore, a range of metrics and terms are used to describe the proposed 'pathways', 'roadmaps', and 'journeys' to improvement. In the case of CO_2, the word 'decarbonisation' is used to describe the progressive reduction in the carbon intensity (and attendant decrease in carbon emissions) of any anthropogenic activity.[1] It is thus a process of continual change rather than a fixed state. Decarbonisation should ultimately lead to a situation of carbon neutrality, in which as much carbon dioxide is removed from the atmosphere as is emitted. This removal could occur in the form of carbon capture or storage. Carbon neutrality is also described as net zero CO_2 emissions and is achieved when 'anthropogenic CO_2 emissions are balanced globally by anthropogenic CO_2 removals over a specified period' (IPCC, 2018). Net zero for all emissions, however, is only achieved 'when anthropogenic emissions of all GHGs are balanced by anthropogenic removals over a specified period' (IPCC, 2018). If a process removes more carbon from the atmosphere than it creates, a 'carbon negative' situation could theoretically be reached.

For a sector as complex, international and interrelated as air transport, even defining the inclusion and exclusion criteria for net zero CO_2 targets is contentious, and the focus of aviation's net zero CO_2 target has been on 'owned' carbon, directly created by one entity, and not the sector's non-CO_2 effects. This has led to accusations of 'greenwashing' when airports have stated their carbon accounting will only include so-called Scope 1 and Scope 2 emissions created directly by the airport operator and will not include Scope 3 emissions from third parties, including airlines and tenant companies.

Achieving net zero CO_2 from international flights by 2050 will require a combination of measures, and IATA's (2021) net zero strategy relies on the sector transitioning to a mix of 65% SAFs, 19% offsets and carbon capture; 13% electric, hybrid aircraft, and hydrogen-powered aircraft; and 3% from infrastructure and operational efficiencies. Achieving ICAO's LTAG of net zero carbon emissions by 2050 will thus require not only technological innovations but also changes in infrastructure, operations and the use of carbon offsetting and carbon capture and storage (IATA, 2024a). Each measure poses its own challenges, and no single intervention is a panacea or suitable for application in all cases. Significantly, the LTAG only refers to emissions from international flights, and given the significant (and in many cases growing) domestic air transport markets in the United States, India, Indonesia, and China, among others, the carbon contribution of these flights should not be overlooked. In the section that follows, we review some of the key technological propositions and developments that are predicted to enable the net zero transition before selected regulatory and policy initiatives are presented.

TECHNOLOGICAL PROPOSITIONS

The technological propositions to help decarbonise aviation essentially fall into two main categories: alternative fuels and energy sources (Pintos et al., 2023) and innovative airframe design and materials (Babuder et al., 2024). In the short-to-medium term, SAFs,[7] including biofuels, and electric, hydrogen, or hybrid propulsion have been presented as potential alternatives to kerosene (Pintos et al., 2023). Of these, SAF is the most mature technology. Following decades of research, the latest generation of SAF is compatible with existing aircraft and fuelling infrastructure and so can be 'dropped in' to the supply chain with minimum disruption (IATA, n.d.). A number of trial flights have already been performed, and investment is being made into plants that can manufacture SAF at scale. Other options – including electrification and hydrogen – are also being explored, however both are at a lower stage of technological readiness (see Table 10.1).

Sustainable Aviation Fuel

Kerosene has been aviation's fuel of choice for decades. Energy dense, relatively light, and formerly inexpensive and readily available, it represented a near-ideal source of power. However, upon combustion, a range of atmospheric pollutants are created, and many of them, including CO_2 and water vapour, are important GHGs. Consequently, more environmentally sustainable alternatives to kerosene have been sought. The first-generation biofuels were derived from biological feedstocks, but concern about land-use impact and productive capacity meant that

Table 10.1. Key Challenges Related to the Adoption of Alternative Fuels and Energy Sources.

	SAF	Electric Propulsion	Hydrogen Propulsion
Airline operations	Cost and availability of supply. No significant disruption to routine operations	Cost. Battery technology limited. Range and performance poorer than existing aircraft. Speed and capacity of recharging limited	Cost and impact on aircraft turnarounds and refuelling
Airport operations	Changes in maintenance regimes	Medium-to-high impact in multiple areas, including refuelling, maintenance, external power connection/disconnection, and luggage/cargo loading/unloading	Significant impact on both refuelling and maintenance
Airport infrastructure	Impact on both the fuel depot and the hydrant refuelling system	Significant effect on a number of existing infrastructural assets, including fuel storage, hydrant refuelling system, maintenance hangars, and, to a lesser extent, aircraft stands	Cost of installing and maintaining new infrastructure. Medium-to-high impact on fuel storage, hydrant refuelling system, maintenance hangars, aprons, and aircraft stands

Source: Derived from Babuder et al. (2024).

later generations were manufactured from a range of more renewable sources including agricultural waste, crop oils, wood residues, algae, domestic refuse, and, most recently, human waste.

Using SAF can provide up to 80% reductions in total CO_2 lifecycle emissions compared to kerosene (IATA, n.d.). There is consensus among key policymakers (including ICAO and IATA, the Federal Aviation Administration (FAA) in the United States, and the EU Aviation safety Agency (EASA)) about the potential of SAF (Johnson & Gonzalez, 2013). According to IATA (n.d.), SAF has three distinctive characteristics: it is (a) sustainable,[3] (b) offers a viable alternative to oil, and (c) is appropriate for use in aircraft. SAF can be produced through various processes and the choice of production process and the base feedstock plays an important role in determining whether the production is truly sustainable and whether the resulting fuels can be produced in sufficient quantities (Chiaramonti et al., 2021).

One of the main advantages of SAF is that it can be used in the existing aircraft fleet without any engine or design conversions (Babuder et al., 2024; Pintos et al., 2023). Nevertheless, appropriate certification is needed, and ICAO (n.d.-b) has approved a maximum 50% blend ratio for use in passenger flights. The world's first trans-Atlantic flight using 100% SAF took place on 19 November 2023 by Gulfstream Aerospace Corporation from the company's headquarters in Savannah in United States to Farnborough airport in the United Kingdom (Gulfstream, 2023). A few days later, on 23 November 2023, Virgin Atlantic flew a Boeing 787 powered by SAF from London Heathrow to New York (Virgin Atlantic, 2023). These flights demonstrated that SAF can be used to power both small and large aircraft. However, it has been suggested that SAF-powered engines may require more frequent maintenance than conventionally fuelled ones (Babuder et al., 2024).

Currently, SAF represents less than 1% of the global aviation fuel mix and production capacity is limited. SAF is also up to five times more expensive than kerosene (World Economic Forum (WEF), 2023). To reach a SAF production target of 475 billion litres annually, production will need to increase 27 times between 2030 and 2050 (WEF, 2023). The fact that SAF production is limited means that measures to encourage and support increased SAF production are required. Mayeres et al. (2023) evaluated the cost of different policies to support the adoption of SAF and argued that imposing SAF mandates, which have been introduced by some countries to try and incentivise uptake, is 5–10 times more expensive compared to other options such as emissions trading schemes (ETSs). There is also concern that different SAF mandates in individual countries may lead to increased emissions as airlines tanker fuel to purchase it in locations where it is cheapest.

According to IATA (n.d.), the advantages of using SAF extend beyond environmental benefits and offer economic advantages for airlines by shielding them from volatile oil prices. Additionally, it can support jobs and stimulate research and investment in 'green' energy production. A further advantage of SAF is that it can potentially be created in a more diverse range of locations and offer enhanced future energy security for airlines and nations alike. At present, however, there are relatively few producers of SAF, and the cost of production remains a disincentive to uptake.

Hydrogen and Hybrid–Electric Technologies

Hydrogen presents another potential energy source which is being actively explored as an alternative to kerosene (IATA, 2019). The advantages of hydrogen include (a) zero carbon emissions at point of use, (b) high energy efficiency, (c) no noise or visual pollution, (d) quick fuelling time, and (e) long shelf life (Kaufmann et al., 2010). Hydrogen is abundant and is 2.5–3 times more energy dense per kilogram than traditional jet fuel (Airbus, n.d.-a; IATA, 2019). However, it requires four times more storage space in aircraft (Verstraete, 2009). Airbus aims to bring a low-carbon commercial aircraft to the market by 2035 (Airbus, n.d.-a). The company's ZEROe project is exploring a variety of configurations and technologies and at the same time is preparing the ecosystem that will produce and supply the hydrogen (Airbus, n.d.-b).

Hydrogen can be used either directly as an alternative to kerosene for large aircraft or converted into electrical power for fuel cells appropriate for small aircraft (Airbus, n.d.-a; IATA, 2019; Kaufmann et al., 2010). The production of hydrogen can be achieved through thermochemical or electrochemical means to separate oxygen and hydrogen from water by using electricity (CSIRO, 2018). Although the use of hydrogen power appears promising in the longer term, key challenges remain. These concern safe and routine hydrogen storage and handling, the need to redesign refuelling procedures, impacts on turnaround operations, incompatibility with existing infrastructure, and the need to install a completely new fuel delivery system (Babuder et al., 2024). However, the biggest challenge is the cost of production (IATA, 2019; Kaufmann et al., 2010; Yusaf et al., 2023).

Hydrogen can also be used to power fuel cells in hybrid-electric propulsion systems. Owing to their lower power density and higher cooling requirements when compared to conventional propulsion systems, hybrid systems are only considered suitable for very small short-range aircraft and are unlikely to power large commercial aircraft (Akkaya et al., 2024). However, research by Thonemann et al. (2024) using a 50-passenger aircraft demonstrates that if we wish a hybrid-electric aircraft to be sustainable, it should fully account for the environmental impact over the entire life cycle from raw material extraction to end-of-life disposal.

Electric Propulsion

A third potential energy source is electricity. Although the idea is not new, it is only relatively recently that developments in battery technology have made short-range flights with limited payloads feasible (IATA, 2019). One company, Eviation, has created a full electric aircraft called 'Alice' (Eviation, n.d.), and other companies are testing hydrogen-electric-powered aircraft (Beyond Aero, 2024; Universal Hydrogen, 2023; ZeroAvia, 2023) or hybrid-electric (Ampaire, 2022; VoltAero, 2024). Researchers have explored the application of all-electric aircraft in specific locations. Eaton et al. (2023) investigated the potential of all-electric flights to decarbonise US aviation, while Pintos et al. (2023) studied the potential use of electric aircraft on inter-islands services in the Canary Islands. Three route flight profiles were analysed, and the fuel consumption and emissions generated by conventional propulsion and sustainable alternatives were compared. The results

commitments on how to achieve net zero emissions by 2050 (Department for Transport, 2022).

Scheelhaase et al. (2012) argued that the cost of the EU ETS for airlines would be affected by the allowances which were issued for free and the number of allowances needed to maintain operations in the future. Pérez-Calderón et al. (2021) analysed the cost of the EU ETS on 10 European airlines during the period 2012–2019 and showed that among those with the highest level of efficiency in emissions (CO_2 per RPK) are the largest in terms of revenue passengers per kilometre (RPK) and available seats per kilometre (ASK)[5] while the airlines with the worst performance in terms of relative CO_2 emissions were selected small airlines and low-cost carriers.[6] On the contrary, Fageda and Teixidó (2022) analysed a sample of intra-EU flights between 2010 and 2016 to examine the causal impact of the EU ETS on emissions and supply. The authors contend that the EU ETS reduced emissions by 4.7% on the regulated routes. What is more, the impact was found even higher (10.7%) in a restricted sample of short-haul flights (less than 1,000 km) in which competition with other means of transport may exist. Also, the reduction in emissions was found to be high for low-cost airlines (−11%) but not statistically significant for full-service carriers.

The EU ETS scheme was opposed by some in the international community and in December 2009, the Air Transport Association of America (ATA) and the US airlines American, Continental and United commenced legal action against aviation's inclusion in the EU ETS (Court of Justice of the European Union, 2011). Opponents started preparing countermeasures against European airlines and China reportedly suspended orders of Airbus aircraft worth US$12 billion (Airbus, 2012, as cited in Scheelhaase et al., 2012). The pressure from the international community and industry meant the policy was only applied to flights within the European Economic Area (EEA) – EU Member States, plus Iceland, Liechtenstein and Norway. As the EU ETS is based on intra-EU flights, 58% of CO_2 emissions were unaccounted for (European Commission, n.d.-a). Global coordinated action was thus required.

In October 2016, ICAO agreed on a global market-based measure to stabilise CO_2 emissions at 2020 levels (ICAO, 2016). Under the scheme, known as CORSIA – Carbon Offsetting and Reduction Scheme for International Aviation, emissions from international flights will be accounted for. In 2026, CORSIA will be evaluated by the European Commission and depending on the findings, a legislative proposal might be introduced for the expansion of the EU ETS to flights departing from the EU to other parts of the world (European Commission, 2022). The phasing out of the free allocation will be coupled with the introduction of SAF allowances (European Commission, 2022). IATA (2023) welcomed the proposal to encourage SAF production.

The international nature of aviation shows the complexity of applying emissions schemes without disproportionately impacting the aviation industry in other countries. Pérez-Calderón et al. (2021) argues that an international commitment to reconcile the EU ETS with other CO_2 emissions monitoring schemes such as CORSIA will not be easy to secure. Maertens et al. (2019) analysed different options to continue the EU ETS after the adoption of CORSIA and showed that any form of continuation of the EU ETS would have positive environmental effects and especially in the early 2020s, when the coverage and environmental impact of CORSIA, which only tackles any post-2020 emissions' growth in international aviation, will still be limited. The authors propose alternative scenarios to keep active both CORSIA and EU ETS (see Maertens et al., 2019): (a) EU ETS could be applied only for intra EU flights or even for international flights within EEA as an alternative to CORSIA, or (b) voluntarily extend CORSIA to domestic EEA flights. Efthymiou and Papatheodorou (2019) likewise concluded that it is possible to link the EU ETS with other ETSs to create a global ETS.

The EU ETS deals with CO_2 emissions; however, researchers argue that the environmental impact of aviation goes beyond CO_2 and includes nitrogen oxides, sulphurous oxides, and unburnt hydrocarbons, carbon monoxide, particulate matter, and water vapour (Modarress & Ansari, 2020) which are at least as harmful for the environment as CO_2 (European Commission, 2020). The European Commission delegated the EASA to conduct an updated analysis about the non-CO_2 effects of aviation on climate (European Commission, 2020). In 2023, a new directive was introduced by the European Commission to provide further steps to tackle non-CO_2 emissions of aviation, and a consultation process took place in 2023 with stakeholders (European Commission, 2023).[7] By 31 August 2024, another reform of the EU ETS is the introduction of a system to monitor, report, and verify (MRV) the airlines' non-CO_2 emissions. Analysis of the data will be performed by the Commission and will evaluate the introduction of a proposal to address non-CO_2 emissions from aviation by 31 December 2027 (European Commission, 2023).

In July 2021, the European Commission published a package of legislative proposals called 'Fit for 55' to make all sectors of the EU's economy reach its climate targets in a fair, cost-effective, and competitive way. Under the European Climate Law, the EU committed to reduce its net GHG emissions by at least 55% by 2030. To further push the use and uptake of SAFs, one component of the package is the 'ReFuelEU' proposal which aims to boost production and uptake of SAF.[8] IATA (2023) welcomed the scheme but emphasised that it is essential to have a comprehensive framework for providing incentives for cheaper production. Furthermore, it emphasised the need for a harmonised system for the creation of a flexible SAF market as the Commission recognises that SAF production will be uneven across Europe.

CONCLUSION

According to the World Economic Forum (2023), achieving aviation's 2050 net zero transformation will require a US$5 trillion CapEx investment, equivalent to approximately US$185 billion a year, 2.4 times higher than current levels of investment. Although there are growing arguments to be made in favour of investing in lower emissions technologies, the funding and support mechanisms need to be in place to incentivise innovation and uptake. As this chapter has shown, while technological propositions, including SAF, may be able to deliver some of the required emission reductions, policy interventions, whether in the form of mandates, economic instruments such as emission trading schemes or, ultimately, perhaps, politically unpalatable demand management interventions, may also be required. What is evident is that the status quo cannot continue. The global temperature records cited at the beginning of this chapter have now been broken again for an 11th consecutive month while ocean temperatures continue to exceed all records. To enable aviation to decarbonise at pace will require considerable financial investment, political will, technological innovation, and concerted stakeholder engagement. Although aviation's decarbonisation journey has begun, the path to net zero will be paved with unanticipated challenges and decisions which are taken now will shape not only aviation's future, but the future of the planet. It is vital, therefore, they are informed and considered.

NOTES

1. While aviation also emits CO (carbon monoxide), it is CO_2 as a GHG which attracts most attention.
2. SAF is the main term used by the aviation industry to describe new fuels which are not derived from fossil fuel. Related terms include sustainable alternative fuel, sustainable alternative jet fuel, renewable jet fuel, or biojet fuel (IATA, n.d.).
3. For SAFs to be sustainable, they should ensure at least a 65% reduction in their carbon intensity against the baseline of 94 g CO_2 equivalent per megajoule (gCO_2 eq/MJ) for fossil fuels (Chiaramonti et al., 2021).
4. At the time of writing, EU ETS is in its fourth phase (2021–2030).
5. The study refers to AirFrance-KLM, IAG, and Lufthansa.
6. The study refers to Ryanair and Aegean.
7. Directive (EU) 2023/958. https://eur-lex.europa.eu/eli/dir/2023/958/oj
8. The RefuelEU aviation regulation will boost SAF use and production through the following ways: (1) The aircraft fuel suppliers at EU airports will increase the share of sustainable fuels that they distribute (2% by 2025, 6% by 2030, 20% by 2035, 34% by 2040, 42% by 2045, and 70% by 2070). (2) Airlines departing from EU airports refuel only with the fuel necessary for the flight and avoid tankering. (3) EU airports guarantee the necessary infrastructure to deliver, store, and refuel SAF (Council of the European Union, 2023).

REFERENCES

Airbus. (n.d.-a). *Hydrogen. An important decarbonisation pathway*. Retrieved March 30, 2024, from https://www.airbus.com/en/innovation/low-carbon-aviation/hydrogen#:~:text=Two%20primary%20uses%20for%20hydrogen%20in%20aviationandtext=Hydrogen%20propulsion%3A%20Hydrogen%20can%20be,gas%20turbine%20via%20fuel%20cells

Airbus. (n.d.-b). *ZEROe Towards the world's first hydrogen-powered commercial aircraft*. Retrieved March 30, 2024, from https://www.airbus.com/en/innovation/low-carbon-aviation/hydrogen/zeroe

Akkaya, M., Neumann, N., & Peitsch, D. (2024). A method for an efficiency and weight-optimised preliminary design of a hydrogen-powered fuel cell-based hybrid-electric propulsion system for aviation purposes. *CEAS Aeronautical Journal, 15*(2), 175–189.

Ampaire. (2022). *Ampaire flies longest ever nonstop hybrid-electric mission*. Retrieved April 4, 2024, from https://www.ampaire.com/_files/ugd/03bbf3_e8cf3dea3e454befb3af62c023f52672.pdf

ATAG. (2020). *Aviation benefits beyond borders*. Retrieved January 6, 2024, from https://aviationbenefits.org/media/167517/aw-oct-final-atag_abbb-2020-publication-digital.pdf

Babuder, D., Lapko, Y., Trucco, P., & Taghavi, R. (2024). Impact of emerging sustainable aircraft technologies on the existing operating ecosystem. *Journal of Air Transport Management, 115*, 102524.

Beyond Aero. (2024). *Introducing France's first manned hydrogen-electric flight. We keep saying that Aviation will be electric. Turns out, it just did.* Retrieved April 30, 2024, from https://www.beyond-aero.com/press

Boston Consulting Group (BCG). (2021). *Plotting aviation's uncharted course to net zero.* Retrieved July 10, 2024, from https://www.bcg.com/publications/2021/net-zero-aviation-is-the-future-of-aviation

Chiaramonti, D., Talluri, G., Vourliotakis, G., Testa, L., Prussi, M., & Scarlat, N. (2021). Can lower carbon aviation fuels (LCAF) really complement sustainable aviation fuel (SAF) towards EU aviation decarbonisation? *Energies, 14*(19), 6430.

Copernicus. (2024). *Copernicus: March 2024 is the tenth month in a row to be the hottest on record.* Retrieved April 9, 2024, from https://climate.copernicus.eu/copernicus-march-2024-tenth-month-row-be-hottest-record

Council of the European Union. (2023). *Fit for 55: Increasing the uptake of greener fuels in the aviation and maritime sectors.* Retrieved January 12, 2024, from https://www.consilium.europa.eu/en/infographics/fit-for-55-refueleu-and-fueleu/

Court of Justice of the European Union. (2011). *Judgement of the Court (Grand Chamber) 21 December 2011 in case C-366/10, Luxembourg.* Retrieved January 9, 2024, from https://eur-lex.europa.eu/legal-content/EN/TXT/?uri=CELEX%3A62010CJ0366

CSIRO. (2018). *The national hydrogen roadmap provides a blueprint for the development of a hydrogen industry in Australia.* Retrieved April 2, 2024, from https://www.csiro.au/en/work-with-us/services/consultancy-strategic-advice-services/csiro-futures/energy-and-resources/national-hydrogen-roadmap

De Bortoli, A., & Féraille, A. (2024). Banning short-haul flights and investing in high-speed railways for a sustainable future? *Transportation Research Part D: Transport and Environment, 128*, 103987.

Department for Transport. (2022). *Jet zero strategy delivering net zero aviation by 2050.* Retrieved January 7, 2024, from https://assets.publishing.service.gov.uk/media/62e931d48fa8f5033896888a/jet-zero-strategy.pdf

Directive (EU) 2023/958. Retrieved February 20, 2024, from https://eur-lex.europa.eu/eli/dir/2023/958/oj

Eaton, J., Naraghi, M., & Boyd, J. G. (2023). Market capabilities and environmental impact of all-electric aircraft. *Transportation Research Part D: Transport and Environment, 124*, 103944.

Efthymiou, M., & Papatheodorou, A. (2019). EU emissions trading scheme in aviation: Policy analysis and suggestions. *Journal of Cleaner Production, 237*, 117734.

European Commission. (2020). *Updated analysis of the non-CO_2 effects of aviation.* Retrieved January 20, 2024, from https://climate.ec.europa.eu/news-your-voice/news/updated-analysis-non-co2-effects-aviation-2020-11-24_en

European Commission. (2022). *European green deal: New rules agreed on applying the EU emissions trading system in the aviation sector: Press release IP/22/7609.* Retrieved January 13, 2024, from https://ec.europa.eu/commission/presscorner/detail/en/ip_22_7609

European Commission. (2023). *Consultation on monitoring, reporting and verification of aviation non-CO2 impact on the climate.* Retrieved January 22, 2024, from https://climate.ec.europa.eu/news-your-voice/news/consultation-monitoring-reporting-verification-aviation-non-co2-impact-climate-2023-10-25_en

European Commission. (n.d.-a). *Development of EU ETS (2005–2020).* Retrieved January 7, 2024, from https://climate.ec.europa.eu/eu-action/eu-emissions-trading-system-eu-ets/development-eu-ets-2005-2020_en

European Commission. (n.d.-b). *EU emissions trading system (EU ETS) – Free allocation – Carbon leakage.* Retrieved January 7, 2024, from https://climate.ec.europa.eu/eu-action/eu-emissions-trading-system-eu-ets/free-allocation/carbon-leakage_en

Eviation. (n.d.). *Alice: Built to make flight sustainable, affordable, quit solution to regional travel.* Retrieved April 4, 2024, from https://www.eviation.com/aircraft/

Fageda, X., & Teixidó, J. J. (2022). Pricing carbon in the aviation sector: Evidence from the European emissions trading system. *Journal of Environmental Economics and Management, 111*, 102591.

Gössling, S., Humpe, A., & Sun, Y. Y. (2025). Are emissions from global air transport significantly underestimated? *Current Issues in Tourism, 28*(5), 695–708.

Gov.uk. (2024). *Guidance participating in the UK ETS* (updated 2 January 2024). Retrieved February 25, 2024, from https://www.gov.uk/government/publications/participating-in-the-uk-ets/participating-in-the-uk-ets

CHAPTER 11

MARITIME TRANSPORT IN A NET ZERO ECONOMY

Jason Monios

Kedge Business School, France

ABSTRACT

This chapter reviews the current state of carbon emissions in the maritime transport sector and discusses the prospects of reaching net zero. It begins with an overview of current fuel usage and emissions in maritime transport, which account for 2.8% of global anthropogenic carbon emissions. The latest study from the International Maritime Organization (IMO) predicts that the current level of emissions is expected to increase by 50% by 2050, with the best-case scenario only able to maintain the current level. This chapter then discusses the alternative fuels and practices that may contribute towards decarbonisation of the sector, such as hydrogen, ammonia, methanol, and wind power, revealing that only a tiny number of vessels have renewable fuel capability, and that is mostly dual-fuel engines that will continue to burn fossil fuels. Next, this chapter evaluates shipping policies and governance mechanisms seeking to drive a trend towards net zero, including the roles of key institutions such as the European Union (EU) and the IMO. This chapter concludes that, despite a proliferation of small-scale industrial trials, there is currently no credible pathway to net zero for maritime transport outside of major policy intervention aligned with significant reduction in demand for goods transport.

Keywords: Decarbonisation; alternative fuels; greenhouse gas; maritime transport policy; shipping governance; climate change

---

Towards Transport Net Zero
Transport and Sustainability, Volume 20, 159–177

Published under exclusive licence by Emerald Publishing Limited
ISSN: 2044-9941/doi:10.1108/S2044-994120250000020011

INTRODUCTION

The maritime transport sector is not yet serious about decarbonisation. Policy targets remain vague regarding a commitment to full decarbonisation by 2050, while there is currently no prospect of replacing the existing fossil-fuelled fleet by this date. Even if this were likely, there is little prospect of the availability of sufficient alternative fuel to power these engines. It is becoming common to describe shipping, like aviation, as a 'harder-to-abate' sector (Energy Transitions Commission (ETC), 2018). Excuses are already in place, and industry representatives expect the sector to be largely fossil fuelled well past 2050, probably until around 2070 (McKinsey, 2023).

The role of maritime transport in a net zero economy is dependent on the available energy mix. If net zero shipping fuels are abundant, then the supply of shipping could meet demand (which in any case is likely to be lower in a net zero economy due to reduced production and consumption of goods). If net zero fuels are not widely available (and carbon fuels are not permitted), then the supply of maritime transport could be much reduced and international trade significantly limited as a result. As this trade limitation is not likely to be politically acceptable, it is difficult to imagine that in this scenario fossil fuels would actually be forbidden. Thus it is likely that, unless and until zero-carbon fuel is widely available, the maritime industry will continue to operate on fossil fuels for decades to come, a view widely held in the industry (Duru, 2019; McKinsey, 2023).

This chapter will outline the current fuel usage of the sector and then review the potential alternatives, evaluating their advantages and disadvantages. As the primary disadvantage is that the major alternative fuels require green electricity as a feedstock, the likely consequences of a limited supply will then be considered. The role of policy in managing this situation will then be discussed before drawing final conclusions on the likely progress towards the decarbonisation of the maritime sector. This progress is dependent on the need for strong policy support such as banning fossil fuels, mandating that the industry channels its efforts into a single alternative (hydrogen), and devoting more resources to wind power, since it is the only fully green alternative.

OVERVIEW OF SHIPPING DEMAND AND SUPPLY, FUELS USED, AND EMISSIONS

The global commercial shipping fleet totals 105,500 vessels over 100 gross tonnes (including 56,500 ships over 1,000 gross tonnes) with a total volume of 2.3 billion dead weight tonnes, triple the level observed in 2000 (UNCTAD, 2023). The primary specialised vessel types are dry bulk carriers (43% of the fleet), oil tankers (29%), and container ships (13%). The remainder of the fleet is made up of general cargo and other vessels, including 0.4% of tonnage represented by passenger vessels and ferries. Despite a lot of focus in society on decarbonisation by reducing consumer demand, it is notable that only 13% of maritime transport capacity is devoted to containerised goods (including but not limited to consumer items). The majority of freight transported is bulk goods that fuel industry, such as iron ore for steel production, coal for energy production, oil and gas for domestic uses such as heating and transport fuel, and grain for animal feed. A corollary of the above is that around 40% of maritime transport consists of fossil fuels (oil, gas, and coal), meaning that in a net zero economy, if the world economy were to decarbonise and such inputs were no longer needed, maritime transport demand would thus shrink by an equivalent amount, which would make the decarbonisation task much smaller. On the other hand, some of this lost demand is likely to be replaced by transportation of alternative fuels such as hydrogen and ammonia.

The future level of demand for shipping is open to speculation. In the short term, UNCTAD (2023) predicts annual growth of 2.1% up to 2028. Long-term demand will be influenced both by global economic growth and geopolitical and other external factors that will likely increase the cost of shipping, as may environmental regulations requiring more expensive fuel or engine adaptations. Consultants including McKinsey (2017) predict continual growth, while other authors caution that future disruptions due to climate change and geopolitics may lead to a reduction (Monios & Wilmsmeier, 2022). The greenhouse gas (GHG) emissions produced by this current level of demand amounted to 1,076m tonnes (in CO_2e) in 2018, 9.6% more than 2012 (977m tonnes), representing 2.89% of total anthropogenic GHG emissions (IMO, 2020).[1] The IMO study predicts a best-case scenario of the same level of carbon emissions in 2050, with a worst-case scenario of up to a 50% increase. These figures depend on the total demand for shipping, the efficiency of the vessel engines and operational practices, and the mix of fuels used.

The IMO GHG study shows the total amount and types of fuel used across the sector. The vast majority (79%) of fuel used is heavy fuel oil (HFO), which used to be refinery waste after the other types of oil were refined, until it was realised that it can be used as fuel itself. It is so thick that it needs to be heated before it can be used in the engine. The remaining 21% is split between marine diesel oil (MDO) and liquefied natural gas (LNG). These fuels differ in their levels of sulphur oxides (SO_x), nitrogen oxides (NO_x), and particulate matter (PM) produced; therefore, they produce less local air pollution and can be used to meet certain clean air regulations such as emission control areas (some control for SO_x only, some cover both SO_x and NO_x). However, they are mostly identical in terms of carbon emissions (as a gas, LNG produces lower carbon emissions than oil when combusted, but from the lifecycle perspective, partly due to methane slip, its overall carbon emissions are similar to or potentially even higher than oil). Table 11.1 presents the emissions profile of the main fossil fuels currently used in shipping. There is currently no significant use of carbon-free alternatives.

new vessels). However, many significant problems remain with regard to ammonia. Not only does it require yet more large use of energy to produce in addition to that used to produce the hydrogen, but it is highly toxic both to humans and marine life and when combusted produces more N_2O (a powerful GHG) than conventional fuel, hence potentially undermining its own carbon emission savings (Tomos et al., 2024). Technologies such as selective catalytic reduction (SCR) or exhaust gas recirculation (EGR) are already used by ships to reduce NO_x emissions when required by emission control areas, but they do not remove them entirely. It is possible that, unless both the toxicity problem due to nitrogen slip and the GHG increases due to N_2O emissions can be resolved soon, ammonia could be abandoned as a possible shipping fuel. As ammonia is included as a key option in many reports and studies (Global Maritime Forum, 2021; Transport and Environment, 2022), this question needs to be answered urgently in order to inform future shipbuilding strategies.

Carbon-neutral fuels are less desirable because they only recycle emissions, but at least they do not produce any new emissions. They can either be biofuels or e-fuels. Biofuels have the advantage in that (in theory but not always in practice) they use existing organic waste from crop waste, food waste, animal waste, sewage, and used cooking oil. Another attractive point is that they can be used as drop-in fuels in engines that already use the conventional non-bio version. Bio-diesel is already widely used in trucks, for example. Bio-methanol is proving particularly popular at the moment, with Maersk announcing its intention to commit large investments to bio-methanol and even establishing their own company to produce it. Maersk has 25 dual-fuel methanol container ships on order, ranging from a small 2,100 TEU to a large 17,000 TEU (Maersk, 2023).

One of the drawbacks of biofuels is in purposely planting crops such as palm oil in order to be turned into biofuels, particularly in developing countries where it may compete with food crops and is one of the major causes of deforestation, and it is difficult to reliably document this sourcing. The other main drawback is that there is already high demand from other sectors (e.g., road transport), and we cannot produce enough of it, so it can never be a serious replacement for the total demand. Even in Maersk's own press release announcing the maiden voyage of their 16,000-TEU methanol-enabled container ship, they state: 'Maersk has secured sufficient green methanol to cover the vessel's maiden voyage and continues to work diligently on 2024–25 sourcing solutions for its methanol-enabled vessel fleet' (Maersk, 2023). Thus, there is the danger that, if we switch to many methanol ships, they are likely to have to use e-methanol (Lloyd's List, 2023) or simply continue using conventional fossil fuel oil (as they are all dual-fuel vessels).

Carbon-neutral e-fuels also have the advantage of being drop-in fuels that can be used in existing engines as they are simply synthetic versions of existing fossil fuels. The main drawback with all e-fuels, as mentioned, is that they require (green) electricity as a feedstock. Moreover, the cost of capturing carbon from the atmosphere is very high. There is also a real problem with inefficiency, as much of the energy is lost translating electrical energy to other fuels, and there is an implicit danger of double counting of carbon reduction emissions from the capture and use, which must be strictly monitored. According to Lindstad et al. (2023), from a full lifecycle perspective, to produce 1 kWh of propulsion even from green electricity to a battery-powered vehicle requires an input of 1.5 kWh of energy, due to losses in transportation and battery usage. When converting that electrical energy into the leading e-fuels, large amounts of energy are lost, meaning that to produce 1 kWh of propulsion of the vessel, the amount of energy put in is 4.2 kWh for e-ammonia, 5.0 kWh for e-hydrogen, 5.9–6.5 kWh for e-methanol, and 6.3–7.1 kWh for e-diesel. Such a high rate of inefficiency questions whether, in a world where energy is scarce, so much can be utilised in converting to e-fuels. The authors calculate the total annual energy demand of shipping as around 5.3 exajoules (EJ). If this were to be provided with e-hydrogen, with an efficiency ratio of only 1:5, it implies that five times this amount of green electricity would need to be produced solely to decarbonise the shipping industry. As a result of this calculation, Lindstad et al. (2023) speculate that, since there will most likely not be enough green electricity to produce this amount of fuel in the decades to come, it is more likely to be directed to sectors where the losses are not so high, e.g., using the electricity directly in battery-powered road transport and domestic uses, with shipping continuing to use fossil fuels until around 2070.

Table 11.3 shows the current world fleet and order book for zero-carbon and carbon-neutral vessels (thus fossil fuels such as LNG are not shown). The first thing to note is the different types of engine. A single-fuel engine obviously uses just that fuel, which can include 'drop-in' fuels, for example, using e-diesel as well as regular diesel in a diesel engine. A dual-fuel vessel is designed to be able to burn more than one type of fuel in its engine. This is the most common way of using LNG at the moment and can provide reliability and flexibility. The third type is called 'ready', for example, ammonia ready. This means that they use conventional fuel oil (usually diesel) but can use a small amount of ammonia mixed

Table 11.3. Current Fleet and Orderbook for Zero-carbon and Carbon-neutral Vessels.

Single Fuel, Dual Fuel, or 'Ready'	Fuel	Vessels in Operation	Percentage of World Fleet	Vessels on Order	Percentage of Order Book
Single	Hydrogen	2	0.0002	3	0.07
	Biofuel	62	0.059	5	0.11
Dual fuel	Conventional/hydrogen	3	0.003	0	0
	Conventional/methanol	23	0.022	62	1.38
	Methanol/biofuel	0	0	2	0.04
Ready	Hydrogen ready	3	0.003	6	0.13
	Ammonia ready	22	0.021	131	2.91
	Methanol ready	0	0	34	0.75
Total		115	0.1082	243	5.39

Source: Adapted from Tomos et al.'s (2024) analysis of data from the Clarkson's shipping register.

Note: While both methanol and LNG can be used either in their fossil fuel or carbon-neutral (biofuel or e-fuel) variants, only methanol is included here as all methanol ships are intended to be fuelled by the carbon-neutral option. All LNG ships are fuelled exclusively by fossil fuel, and they will not use any carbon-neutral version until at least 2040 (see policy discussion); therefore in this chapter, LNG is not considered a genuine alternative fuel.

in. The engine components have been upgraded and modified to be able to deal with the corrosiveness of ammonia and with future full conversion in mind. The table shows that decarbonised fuels are barely evident among either existing or ordered vessels. Only two hydrogen vessels exist with three on order. All methanol ships are dual fuel. There is a clear trend towards 'ready' vessels which may never use any of these fuels or may not use green versions. There is also a clear trend towards ammonia and methanol.

It should also be borne in mind that vessel engines can be retrofitted, thus not all alternatively fuelled vessels in future will be newbuilds. The Global Maritime Forum (2021) proposed a 2050 scenario where approximately half of zero-emission vessels would be newbuilds and half retrofits of existing vessels. According to Lagemann et al. (2023):

> From a shipowner perspective, both methanol and LNG appear to be relatively robust initial power system choices for a broad range of emission reduction ambitions. Both power systems enable low-cost fuel switches to bio- or electro-fuels, while LNG potentially also enables low-cost retrofits to, e.g., ammonia. These findings coincide with the current orderbook indicating that LNG and lately methanol engines are preferred options by shipowners.

The Danger of Lock-in to Fossil Fuels

There is a significant danger that the proposed fuel options encourage lock-in to existing fossil fuels, due to three reasons: drop-in use of e-fuels, the use of dual-fuel engines, and the production of zero-carbon fuels from fossil sources.

The reliance on the availability of drop-in e-fuels makes it more likely that the industry will carry on longer with fossil fuel engines because they can just use e-versions, or they can just say that they will switch to e-versions as they are available and cost competitive. This is what is happening with the EU's proposed ban on the sale of new internal combustion engine (ICE) trucks from 2035, which has now been relaxed to allow such vehicles to be sold as long as they only use e-fuels in the engines. This will delay adaptation as we will continue to produce the same engines, and when (as is likely) sufficient quantities of the e-versions of these fuels are not available (or are not cost competitive), we will continue burning fossil fuels (unless it is prevented through policy).

The second driver of lock-in to fossil fuels is the use of dual-fuel ships. The majority of existing LNG ships, for example, are actually dual fuel, which means they can use fuel oil when LNG is not available (see Lindstad et al., 2021, for more details on e-fuels and dual-fuel engines). Despite a slow increase in LNG ships in contrast to the hype during the last two decades (Monios & Fedi, 2023), and the World Bank recently suggesting that the potential for LNG as an alternative fuel has peaked (Englert et al., 2021), popularity of LNG is suddenly growing; 941 vessels (0.89%) of the world fleet are currently dual-fuel LNG/conventional, with 871 on order, comprising 19% of the order book (Tomos et al., 2024). This change is due both to the Russian invasion of Ukraine leading to gas-importing countries needing to diversify their supply from around the world (many of the LNG-fuelled ships are actually LNG carriers) and to recent tightening of regulations around carbon intensity, which can easily be met by switching to LNG (see discussion in policy section). As shown in Table 11.3, all methanol ships on order are dual fuel or methanol ready; none are 100% methanol only. The dual-fuel option makes the transition smoother; for example, a dual-fuel ship could be able to use either methanol or diesel, begin with the latter and then simply increase its use of the former as it becomes available. The problem here is that it also means that the vessel can simply continue to use fossil fuels forever if the new fuels do not become available in sufficient quantity or at a suitable price.

Finally, another danger, even of preparing for green-electricity-derived hydrogen and ammonia, is that they are currently mostly produced from fossil fuels (so-called blue or grey as opposed to green). Thus, if we switch the fleet to engines that can use hydrogen or ammonia, this could incentivise continued production of such fuels since the green versions produced from electricity will not likely be available in sufficient quantities or at attractive prices. As such, we would only swap the current fossil fuels for new fuels produced (less efficiently) from fossil fuel.

OPERATIONAL METHODS: EFFICIENCES, SLOW STEAMING, AND WIND ASSISTANCE

A great deal of research focuses on practices to reduce energy consumption, other than switching fuels (e.g., Bouman et al., 2017; CE Delft, 2023; Psaraftis & Lagouvardou, 2023). Bouman et al. (2017) calculated that, if a whole suite of options were applied across the sector, about 75% of emissions could be saved by 2050. The measures included alternative fuels (favouring biofuels and LNG, which shows how quickly things have moved on since then), operational measures (mostly slow steaming), alternative energy (primarily wind assistance), hull design (mostly efficiency through ship size), and power and propulsion systems (e.g., hybrid power and waste heat recovery). A more recent report by CE Delft (2023) calculated that the shipping sector could reduce emissions by 28%–47% by 2030 based on 2008 levels through 20%–30% speed reduction and other operational measures (accounting for 50% of the emission reduction), 5%–10% of energy from zero-carbon fuels (in this case, ammonia in dual-fuel engines) (25% of the reduction), and the use of wind assistance technologies (the remaining 25% reduction). They add that applying such measures would only increase costs by around 6%–14%.

One of the most promising and simplest of these options is reducing the speed at which the vessels sail (known as ‘slow steaming’), because the fuel use (hence emissions) from shipping increases cubically as the speed increases linearly. This method is probably the easiest and simplest way to save significant emissions in the short term, although it needs to be planned carefully, for example, with different limits for different ship types and sizes and with sensitivity to weather conditions and route profiles (for more detailed analysis, see Psaraftis, 2019a). This tactic was adopted by shipping lines for economic reasons during the market downturn after the 2009 global financial crisis, as using less fuel saves significant amounts of money. Shippers would need to hold higher inventory due to larger gaps between calls (unless carriers introduced more vessels on a route), and rates

Environment Protection Committee (MEPC) 74 in May 2019, with support from countries such as Greece, but opposition from the UK meant it was ultimately rejected (Psaraftis, 2019b).[12] Market-based measures (MBMs), primarily either a carbon tax or an emission trading scheme (ETS), have been proposed and debated during many years but have never been adopted (Monios & Ng, 2021; Psaraftis, 2012). The process started with a formal request for submissions at MEPC 59 (July 2009) until discussions were formally suspended at MEPC 65 in May 2013. They returned to the table at MEPC 76 in June 2021 but were again postponed pending further work and analysis outside the MEPC.

Research shows the pros and cons of both a carbon tax and an ETS (Lagouvardou et al., 2020; Psaraftis, 2019b). A carbon tax is much simpler than an ETS, although it would be politically difficult to set a high price and difficult to manage at a global level in terms of collecting and distributing the revenues, and the unknown price would not allow any certain achievement of carbon reductions, as the increased cost may simply be absorbed and have no influence on action. While in theory the ETS is better because the amount of carbon emitted can be fixed and the price left to fluctuate according to the market, it comes with administrative complexity and potential offsetting (if it is allowed) rather than reduction of shipping emissions.

The challenges at the IMO are well known, being a forum for decision-making between 174 member states, some of which have more concern than others for their shipping fleets or for their economic reliance on shipping. Some large developing countries, namely China, Brazil, and India, oppose several initiatives, as they believe that their economies would be unfairly disadvantaged by increased maritime costs as a result of regulation. They justify their position according to the principle of common but differentiated responsibilities (CBDR) enshrined in the UNFCCC which, however, conflicts with the IMO policy of 'no more favourable treatment' (Monios & Ng, 2021; Psarafatis & Kontovas, 2020).

Many member states are represented at the IMO by industry, thus the preferred solutions at the IMO tend to be technical (Bach & Hansen, 2023) and those that do not disrupt current practice rather than forceful policy changes (e.g., focusing on increasing efficiency rather than forcing the change of fuels or banning of fossil fuels). There is frequently a gap between the environmental policies of a country towards domestic decarbonisation, and what their representatives lobby for behind closed doors at the IMO, given that they have little incentive to push for decarbonisation of a sector that lies outside their own nationally determined contributions. Monios and Fedi (2023) argue that countries play a dual role in maritime policymaking; first, as agents of the IMO (and the same applies at the EU) which must implement their directives and regulations, and second, as principals who determine the policies and measures adopted by these international bodies. While it is important to identify the inability of a primarily technical organisation such as the IMO to set stringent environmental regulations and policies, attention must also be paid to the role of the member states who lobby against their agreement, delay their implementation, or later refuse to ratify them or neglect to transpose them into national law or enforce them.

Lack of action at the IMO does not preclude individual countries or blocs from taking their own action. The EU has often implemented policies which were later adopted at the IMO level (e.g., regulations on double hulls to prevent oil spills, Port State Control to inspect old vessels, and SO_x limits at EU ports). The EU's carbon ambitions, known as the *Fit for 55* package, are stated as follows:

> In September 2020, the Commission adopted a proposal to cut greenhouse gas emissions by at least 55% by 2030 and put the EU on a responsible path to becoming climate neutral by 2050. To achieve climate neutrality, a 90% reduction in transport emissions is needed by 2050. (European Commission, 2021b)

The two key pieces of legislation adopted by the European Commission are the inclusion of maritime transport in the ETS and the 2023 Fuel EU regulation.

As the IMO has never been able to implement an ETS for maritime transport, the EU has decided to take action. The EU ETS has existed since 2003, but since 1 January 2024, maritime transport has been included. The ETS will include 100% of CO_2 emissions from ships sailing between EU ports, and 50% of emissions between EU and non-EU ports. Methane (CH_4) and nitrous oxide (N_2O) will be added from 2026. The amount of these emissions needing to be covered by ETS permits will be phased in gradually: 40% of 2024 emissions, 70% of 2025 emissions, and 100% of 2026 emissions (European Commission, 2024). Shipping lines are already charging levies for this, e.g., CMA CGM charges €25 per TEU for the Asia to North European route, noting that this rate is based on the early 2024 EU carbon price of €80 per tonne of CO_2. There will be issues of leakage and thus opportunities for neighbouring countries; for example, ships coming from Asia to Europe could tranship their containers at a nearby non-EU port (e.g., in Turkey), thus drastically shortening their non-EU to EU port journey for which 50% of emissions must be accounted. All regional carbon policies face such risks, which over time could perhaps be addressed by use of a carbon border adjustment mechanism (CBAM).

Another important recent EU policy is the Fuel EU regulation. This regulation aims:

> to increase consistent use of renewable and low-carbon fuels and substitute sources of energy in maritime transport across the Union, in line with the objective of reaching Union-wide climate neutrality at the latest by 2050, while ensuring the smooth operation of maritime transport, creating regulatory certainty for the uptake of renewable and low-carbon fuels and sustainable technologies and avoiding distortions in the internal market. (European Parliament, 2023)

Like the ETS, this regulation covers 100% of emissions between EU ports, 50% of emissions between an EU and non-EU port,[13] and also includes emissions while at berth (moreover, the regulation mandates that passenger ships and container ships use onshore power supply while berthed at major EU ports from 2030). The specific limits on well-to-wake GHG intensity of the fuel used by ships (measured in gram of CO_2eq per MJ of fuel) will be scaled up over time, from a reduction on 2020 levels of 2% in 2025 to 80% in 2050 (Table 11.4).[14]

The advantage of this policy is that it cannot be met simply by increasing efficiency or reducing fuel use through slow steaming. The carbon intensity of the actual fuel used must be reduced. This can be done either by adding drop-in fuels

Table 11.4. Fuel EU Regulation GHG Intensity Limits 2025–2050.

Date	Reduction in Carbon Emissions Compared to 2020 (%)
1 January 2025	2
1 January 2030	6
1 January 2035	14.5
1 January 2040	31
1 January 2045	62
1 January 2050	80

Source: Author, based on European Parliament (2023).

to the existing fuel mix or by using a lower carbon fuel in a dual-fuel engine for part of the time such that the overall mix of fuel used in a year meets the reduced target. An additional measure to accelerate uptake of zero-carbon fuels is that, until 2033, the carbon intensity value of renewable fuels of non-biological origin (RFNBO, i.e., e-fuels) will be halved. Another interesting part of the regulation is that it allows the voluntary pooling of emissions between a group of ships, whereby the pool would need to meet the GHG limits on average rather than individually. Research shows, however, that the first three limits can be met by fossil gas such as LNG; thus, it is not until 2040 that genuinely renewable fuels would need to start being used, which runs the risk of delaying transition and locking in fossil fuels (Christodoulou & Cullinane, 2022; Transport and Environment, 2022). Transport and Environment (2022) argues that the policy should not be technology neutral and should specify fixed percentages of renewable fuels required in order to avoid lock-in and start the transition sooner.

In the future, we may see more policy action like this, outside of the IMO, due to fragmentation and delay (Monios & Ng, 2021). Some authors have proposed returning emissions responsibility to nations (Selin et al., 2021). While there is some question of how to apportion these emissions (e.g., by country of ownership of shipping lines or ports) and the likelihood that some countries will continue to take little or no action, other nations could at least take action without waiting for the IMO.

CONCLUSION

The maritime industry does not genuinely expect to be net zero by 2050. The IMO aim was only 50% until recently, and a recent survey of industry stakeholders on the fuel mix in 2050 reveals an expectation that the main fuels (around 50%) will be conventional fuel oil, green ammonia, and bio-diesel, with the rest made up of blue ammonia, LNG, e-methanol, bio-methanol, bio-LNG, and e-LNG (McKinsey, 2023). A dangerous narrative has taken hold characterising shipping as 'harder-to-abate', and analysis of availability of green electricity suggests that shipping (if it continues with a similar level of demand) will continue to use primarily fossil fuels until around 2070 (Lindstad et al., 2023). It is impossible to avoid the conclusion that the maritime industry is not serious about decarbonisation. Only very few authors have raised the possibility that the total volume of shipping will need to reduce in a world where green energy is scarce and expensive (De Beukelaer, 2022; Monios & Wilmsmeier, 2022; Van Leeuwen & Monios, 2022).

There is, as yet, no trend towards solely renewable-powered vessels. All the vessels that can use green fuels are either dual fuel or green fuel 'ready', meaning both could potentially continue using only fossil fuels forever or mixed with a tiny amount of extremely inefficiently produced carbon-neutral e-fuels. These trends raise a serious danger of lock-in to fossil fuel. The Fuel EU regulation is a start but will not bite until 2040 at the earliest. Some authors have proposed banning the use of fossil fuel in shipping by 2050 (Van Leeuwen & Monios, 2022), which would set a fixed deadline and reduce incentives for continued investments in fossil fuels and help to avoid lock-in. A truly net zero society cannot function unless fossil fuels are banned, starting domestically and then considering emissions such as the maritime (and aviation) sector that are currently placed outside nationally determined contributions. In addition, no credible forecast suggests that we will have sufficient green energy to produce the zero-carbon or carbon-neutral fuel for the current level of shipping, thus degrowth scenarios urgently need to be considered.

It is not realistic to expect individual shipowners to invest in green ships if their competitors are free to continue using fossil fuels. Moreover, it is not simply a matter of identifying the best method of decarbonisation in an ideal scenario; our choices will be shaped by the urgent timeline and the scarcity of green electricity as the main feedstock for alternative fuels. Therefore, the following policy steps should be considered for the decarbonisation of shipping in line with a net zero economy by 2050:

- *Efficiency improvements* such as engine and hull designs are already part of IMO policy, and industry will continue to adopt such measures in order to save fuel. This is an essential beginning, but these savings will be overtaken by growth of demand (at least in the short term and in the absence of degrowth scenarios).
- *Slow steaming* is one of the best options we have, to easily and quickly reduce 10%–30% of emissions. Therefore, despite its shortcomings, policymakers could achieve an easy win by mandating speed limits.
- It is time for the industry to get serious about *wind assistance* which is the only genuinely green propulsion option. It can save 20+% in existing large ships, and we should also incentivise fully wind-powered smaller vessels. Policymakers need to mandate the use of wind assistance and do everything possible to expand its use to the entire fleet.
- *We cannot lock-in fossil fuels* by lack of action that allows shipowners to keep using them and intend to keep doing so. Therefore, we must ban fossil fuels by 2050 to send a strong signal that there is no point avoiding renewables. Policymakers should implement an aggressive timetable to phase them out, which will reward early movers.
- *Technology neutral policies are attractive in the short term but will fail in the long term.* Ammonia is toxic and produces too much N_2O which is a powerful GHG; biofuels don't scale; and 'drop-in' e-fuels are inefficient and will be unaffordable due to scarce green electricity that is better used for producing hydrogen.

Global Maritime Forum. (2021). *A strategy for the transition to zero emission shipping*. Retrieved December 7, 2024, from https://globalmaritimeforum.org/report/a-strategy-for-the-transition-to-zero-emission-shipping/

International Maritime Organization. (2020). *Fourth IMO GHG study 2020*. IMO.

International Maritime Organization. (2023) *Resolution MEPC.377(80): 2023 IMO Strategy on reduction of GHG emissions from ships*. IMO.

Lagemann, B., Lagouvardou, S., Lindstad, E., Fagerholt, K., Psaraftis, H., & Erikstad, S. O. (2023). Optimal ship lifetime fuel and power system selection under uncertainty. *Transportation Research Part D: Transport and Environment*, *119*, 103748.

Lagouvardou, S., Psaraftis, H. N., & Zis, T. (2020). A literature survey on market-based measures for the decarbonization of shipping. *Sustainability*, *12*, 3953.

Lindstad, E., Ask, T. O., Cariou, P., & Eskelend, G. S. (2023) Wise use of renewable energy in transport. *Transportation Research Part D: Transport and Environment*, *119*, 103713.

Lindstad, E., Lagemann, B., Rialland, A., Gamlem, G. M., & Valland, A. (2021). Reduction of maritime GHG emissions and the potential role of E-fuels. *Transportation Research Part D*, *101*, 103075.

Lindstad, H. E., Eskeland, G. S., & Rialland, A. (2017). Batteries in offshore support vessels – Pollution, climate impact and economics. *Transportation Research Part D: Transport and Environment*, *50*, 409–417.

Lloyd's List. (2023, September 15). *What is green about methanol?* Lloyd's List. https://lloydslist.com/LL1146620/What-is-green-about-methanol

Maersk. (2023, December 7). *Maersk to deploy first large methanol-enabled vessel on Asia – Europe trade lane*. https://www.maersk.com/news/articles/2023/12/07/maersk-to-deploy-first-large-methanol-enabled-vessel-on-asia-europe-trade-lane

McKinsey. (2017). *Container shipping: The next 50 years*. https://www.mckinsey.com/industries/travel-transport-and-logistics/our-insights/how-container-shipping-could-reinvent-itself-for-the-digital-age

McKinsey. (2023). *The shipping industry's fuel choices on the path to net zero*. https://www.mckinsey.com/industries/travel-logistics-and-infrastructure/our-insights/charting-fuel-choices-as-the-shipping-industry-sails-toward-net-zero#/

Monios, J., & Fedi, L. (2023). The principal-agent problem in hierarchical policy making: A failure of policy to support the transition to LNG as an alternative shipping fuel. *Marine Policy*, *157*, 105846.

Monios, J., & Ng, A. K. Y. (2021). Competing institutional logics and institutional erosion in environmental governance of maritime transport. *Journal of Transport Geography*, *94*, 103114.

Monios, J., & Wilmsmeier, G. (2022). Maritime governance after COVID-19: How responses to market developments and environmental challenges lead towards degrowth. *Maritime Economics & Logistics*, *24*(4), 699–722.

OECD. (2023). *OECD statistics*. https://stats.oecd.org/Index.aspx?DataSetCode=MTE#

Psaraftis, H. N. (2012). Market based measures for greenhouse gas emissions from ships: A review. *WMU Journal of Maritime Affairs*, *11*, 211–232.

Psaraftis, H. N. (2019a). Speed optimization versus speed reduction: Are speed limits better than a bunker levy? *Maritime Economics & Logistics*, *21*(4), 524–542.

Psaraftis, H. N. (2019b). Decarbonization of maritime transport: To be or not to be? *Maritime Economics & Logistics*, *21*(3), 353–371.

Psaraftis, H. N., & Kontovas, C. A. (2020). Influence and transparency at the IMO: The name of the game. *Maritime Economics & Logistics*, *22*, 151–172.

Psaraftis, H. N., & Lagouvardou, S. (2023). Ship speed vs power or fuel consumption: Are laws of physics still valid? Regression analysis pitfalls and misguided policy implications. *Cleaner Logistics and Supply Chain*, *7*, 100111.

Schuler. (2023, July 7). *IMO's revised GHG strategy for shipping: Details and reactions*. GCaptain. https://gcaptain.com/imos-revised-ghg-strategy-for-shipping-details-and-reactions/

Selin, H., Zhang, Y., Dunn, R., Selin, N. E., & Lau, A. K. (2021). Mitigation of CO_2 emissions from international shipping through national allocation. *Environmental Research Letters*, *16*(4), 045009.

Tomos, B. A. D., Stamford, L., Welfle, A., & Larkin, A. (2024). Decarbonising international shipping – A life cycle perspective on alternative fuel options. *Energy Conversion and Management*, *299*, 117848.

Transport and Environment. (2022). *FuelEU maritime: T&E analysis and recommendations. How to drive the uptake of sustainable fuels in shipping*. Transport and Environment.

UNCTAD. (2023). *Review of maritime transport*. UNCTAD.

Van Leeuwen, J., & Monios, J. (2022). Decarbonisation of the shipping sector – Time to ban fossil fuels? *Marine Policy*, *146*, 105310.

CHAPTER 12

MODELLING THE DECARBONISATION AND DIGITALISATION TRANSITION IN LOGISTICS SYSTEMS

Jonathan Köhler[a], Clemens Brauer[a] and Lóránt Tavasszy[b]

[a]*Fraunhofer ISI, Germany*
[b]*Delft University of Technology, The Netherlands*

ABSTRACT

The most important development in shaping the future of logistics is digitalisation, but while digitalisation and decarbonisation are extensively discussed in industry and scientific literature, there are no comprehensive visions of a digitalised, decarbonised logistics system. This chapter addresses the challenges of this transition management process, by framing and illustrating it with quantified scenarios for the future. As a starting point for the autonomous evolution of logistics systems, i.e., without considering politically driven decarbonisation targets, we take the vision of the physical internet. We critically assess this vision with respect to its potential to autonomously bring about the desired levels of decarbonisation. The remaining gap towards targets needs to be actively managed. We develop a transition scenario to complement the physical internet with decarbonisation incentives and assess their combined deployment in the freight system using a quantitative model that simulates transitions in the logistics system.

Towards Transport Net Zero
Transport and Sustainability, Volume 20, 179–194

Published under exclusive licence by Emerald Publishing Limited
ISSN: 2044-9941/doi:10.1108/S2044-994120250000020012

problem of the complexity of managing system-level changes, inspired by notions of institutional and evolutionary economics (Geels, 2012). A key understanding from the transitions literature on sustainable innovation is that system change is not linear. Rather, it is a co-evolutionary process between all the elements of the logistics 'socio-technical' (in the language of transitions thinking) system.

Rather than delving into the detailed components of the socio-technical system of logistics, we investigate the emerging properties of the system, in the sense of its main driving technologies and supporting policies, and how these are adopted within the system. In addition, we focus on quantitative, model-based predictions of how the system performs, to understand which logistics systems and technologies could be successful and how much time transitions between these could take. In this chapter, we present a model that can portray the co-evolutionary process between groups of decision-making agents, and technologies. The model directly predicts policy success in terms of the achieved system-level change within national freight transport systems.

The remainder of this chapter is structured as follows: The next section introduces the MATISSE model, with the agent-based architecture connected to the multilevel perspective of transition management. The third section presents different scenarios for technological futures in the logistics system, ranging from the current approach of Third-Party Logistics (3PL) service provider models to platform-based, synchromodal, and physical internet-inspired systems. Following this, we present the results of the modelling with MATISSE and discuss the implications of this exercise for logistics decarbonisation policy. The final section concludes this chapter with a summary of the findings and recommendations.

SIMULATING TRANSITIONS IN LOGISTICS

Limitations of Current Models

Recent reviews of freight transport analysis methods and models have identified several challenges that follow from the transformation of freight transport and logistics. Meersman and van de Voorde (2019) point out that current transport models have not been successful in predicting the effect of structural, organisational, and behavioural changes on transport demand, and this is one reason why there has been limited success in reducing negative externalities from freight transport. Tavasszy (2020) identifies three challenges for descriptive and predictive transport models. The new structures have to be represented, improved analyses of decision-making in logistics are needed, and enhanced capabilities for analysis of dynamics in freight models should be developed. However, there is no literature that explores the new policy questions that these changes generate. More recently, Köhler and Brauer (2022) have addressed this question. They argue that current freight transport policy models do not address the transitions processes required for systemic change and do not consider the 'twin transition' of combined digitalisation and decarbonisation.

System transitions are hard to model as they are emergent phenomena, a result of many interactions inside the system which have no direct relationship with landscape-level objectives. Because of many interdependent, time-dependent, and often random behaviour of parts of the system, outcomes of change in such systems are difficult to assess. Transition models link the niche level of innovations (e.g., electric vehicles) with regime level (e.g., industry support) and landscape level (overall emissions) changes, so that some innovations can fail and others succeed. These models recognise time dependence, needed as the targets are time-bound and also important as many parts of the system show inertia or have other dynamic responses.

From among the many different possible techniques, we have seen system dynamics and agent-based simulation as the most important approaches. Perhaps the most famous model is the world model used for the 1972 Limits to Growth publication of the Club of Rome (Meadows & Randers, 2013). It predicted, two decades before the Kaya equation was presented, that the world could not sustain continuing growth of population and economies. That marked the start of the sustainability movement, however, this model did not lend itself very well as a description of the transition challenge, as it lacked the levers for transition management. Since then, however, our insights have improved about how to describe such complex and dynamic systems that we wish to manage. Besides system dynamics models being developed further, agent-based models (ABMs) came to the forefront which not only allowed an explicit description of transition management levels but also had a much better operationalisation of the lowest levels, in particular individual decisions, so that policy levers and agent behaviour could be added (Holtz et al., 2015; Köhler & Holtz, 2019).

The remainder of this section introduces a very recent implementation of such a model for the freight transport system and presents an application to demonstrate its importance for decarbonisation policy.

The MATISSE Model

This model portrays how new logistics systems might be adopted in the logistics industry. It defines dimensions to describe in general terms the characteristics of logistics systems. New or alternative logistics systems structures were identified from discussions with the industry and a literature review (Köhler & Brauer, 2022). The adoption/investment decisions of logistics companies about which system they use are then modelled in an ABM framework using a multi-criteria decision rule. The model simulates the time paths of the adoption, to illustrate how the current system might be replaced by new systems. The model is a further development of the approach described in Köhler et al. (2009). Its conceptual structure is described in detail in Haxeltine et al. (2008), while Bergman et al. (2008) provide detail on the equations for the original MATISSE model. The MATISSE model implements the multi-level perspective on transitions using an ABM approach (Köhler et al., 2018), as illustrated in Fig. 12.1. It has two types of agents. There is a small number of logistics system agents which represent the current (regime) and alternative (niche) logistics systems and their characteristics. The other agents are logistics companies who can choose to change the system that they use. There can be only one regime at any time, although the

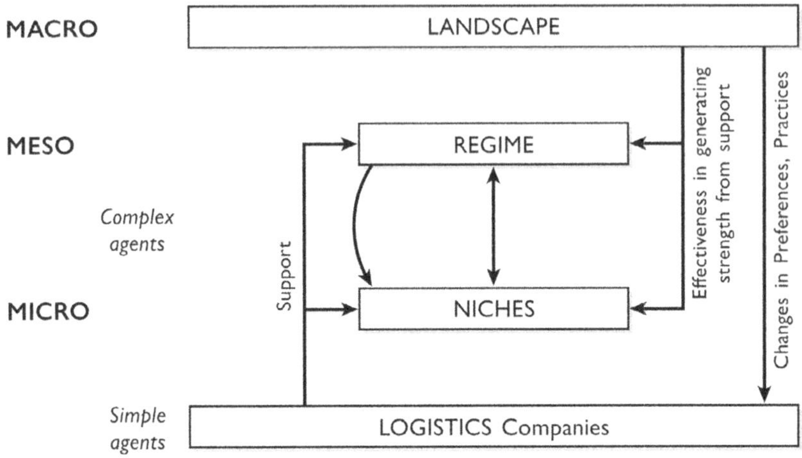

Fig. 12.1. Structure of the MATISSE Model. *Source*: Authors.

system might have periods in which there is no regime. There may be one or several niches at once. The regime is by definition the system which is adopted by the largest number of operators and dominates the system, while niches are much smaller. A logistics system agent's level of adoption determines its behaviour (or strategy), its interactions with other logistics systems, and (partly) its attractiveness to logistics companies deciding which system to adopt.

The model uses the concept of practices as the metric through which agents position themselves in society and over which behaviour is defined. Practices are characteristics of the logistics systems and are at the same time the dimensions over which logistics companies choose the logistics systems they will adopt. Practices are each represented as values along axes, constituting a multidimensional practice space. Agents are differentiated by their positions in the multidimensional practice space. Fig. 12.2 shows schematically a two-dimensional practice space, which might be CO_2 emissions in the x-axis (P_x) and cost of transport in the y-axis (P_y).

The logistics systems and the logistics companies are shown separately for clarity but actually occupy positions along the same P_x and P_y practice axes. Logistics companies are points in the space, while in the figure the size of the regime and niche system ovals are proportional to their relative support. The model is stochastic in that the logistics companies are initially randomly assigned over the practice space, with distributions of conventional logistics companies who use paper management systems and conventional trucks and railways and a smaller number of digitalised logistics companies operating alternative management systems and technologies. Companies are initially assigned in groups according to the system they already use (current logistics, physical internet, synchromodality, regionalised supply chain) over the practice space, with a stochastic distribution of the companies in a group around a central value for the group. In the model,

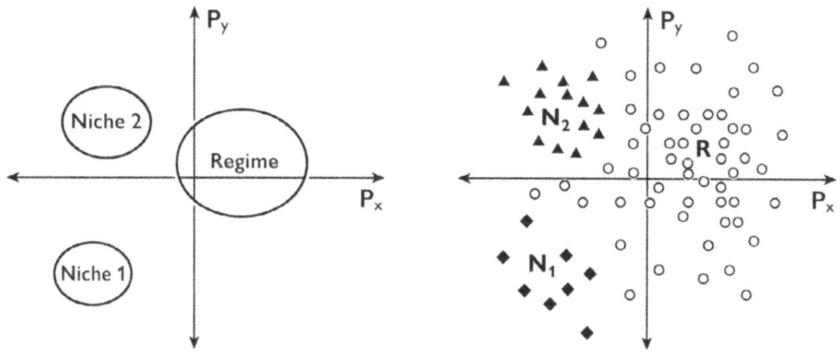

Fig. 12.2. Illustration of a Two-dimensional Practice Space. *Source*: Authors. *Note*: Practice axes P_x and P_y. Left: regime and niche logistics systems, which can move in the practice dimensions space and interact with each other. Right: the logistics companies scattered in the practice space, denoted by the type of logistics system they adopt, ○ = regime (R), ♦ = niche 1 (N_1), and ▲ = niche 2 (N_2).

the logistics agents choose to adopt the practices (logistics system characteristics) of the system that is closest to them; therefore, the positions of the regime and niches are based on clusters of support, i.e., like-minded logistics companies. The positions of the logistics companies in the practice space change depending on landscape signals (exogenous changes in the logistics companies desired system dimensions or practices), so the regime and niche systems develop, i.e., change their characteristics, not only to grow but often just to maintain their adoption by logistics companies.

The dynamics in the model, i.e., the development strategies of the regime and niche logistics systems, were modelled as follows. First, a review of developments in logistics (Köhler & Brauer, 2022) was used to identify the practices of driving factors in logistics companies' adoption of technologies and investment decisions, in a broader way than techno-economic models of freight transport. The dimensions describing the different logistics systems represented in this pilot model (practices in Fig. 12.2) are:

- CO_2 emissions over the supply chain;
- Transport costs along the supply chain;
- The capacity, i.e., the level of activity (tkm) growth or decline in the logistics sector (in the pilot version for the European Union (EU));
- The goal for time to deliver goods to customers;
- Level of digitalisation in data handling, contracts, and documentation for consignments;
- Level of automation of vehicles, loading and unloading equipment, terminals, and warehouses;
- The level of resilience of the supply chain;
- The level of centralisation or concentration of logistics markets and companies.

also enable the decentralisation of terminals if smaller consignment sizes – pallets or parcels – can be combined into the volume and weight envelope of the current standard container. Continuous data updating supports predictive production and distribution, in accordance with statistical or AI-based forecasts of demand in quantity and geographical distribution. This enables a large-scale reduction in warehousing, with the possibility of using the transport vehicle as a short-term warehouse that is mobile and can be rerouted to meet short-term adaptions to the predicted demand based on real-time sales data.

The above three logistics systems could also be configured to explicitly include reduction in emissions as a major feature of their operations. While significant emission reduction is always a possible positive effect of these systems, it is not guaranteed. Additions to secure this decarbonisation would be needed, such as alternative power trains, energy infrastructures, or supporting government policies. Therefore, the research also considered combined digitalised and decarbonised logistics systems: *sustainable logistics control towers, sustainable synchromodal logistics,* and *sustainable physical internet logistics*. These different systems were represented through different initial values of the dimensions (practices) listed above, describing the different logistics systems. In addition to these initial values, the logistics companies change their requirements for performance over time. Scenarios are developed by altering rates of change of the requirements for these dimensions or practices. These alterations may depict policies such as carbon taxes increasing the requirement to reduce emissions or a requirement of logistics companies for even shorter delivery times in response to consumer demand. These rates of change for the baseline and scenarios are shown in Table 12.1. The logistics systems change in response to the changes in the logistics companies' requirements and policies as explained above.

Table 12.1. Rates of Change of Practices.

Initial Rates of Change for Logistics Operators and Logistics Systems								
Name	GHG	Cost	Traffic	Time	Data	Auto	Resilience	Central
Baseline								
3PL	0.0	0.1	0.1	0.0	0.0	0.0	0.0	0.0
Tower	0.0	−0.1	0.1	0.0	0.2	0.0	0.1	0.5
Synchro	0.0	−0.1	0.2	−0.1	0.2	0.5	0.1	0.5
Physical internet (PI)	0	−0.2	0.2	−0.1	0.3	0.5	0.1	0.5
Sustainable tower	−0.5	−0.1	0.1	0.1	0.2	0	0.1	0.5
Sustainable synchro	−0.7	−0.5	0.2	−0.1	0.2	0.5	0.1	−0.1
Sustainable PI	−0.7	−0.5	0.2	−0.2	0.2	0.5	0.1	−0.5
Scenario 1: Initial rates of change for logistics operators								
3PL	−0.5	0.1	0.1	0.0	0.5	0.5	0.0	0.0
Tower	0.0	−0.1	0.1	0.0	0.2	0.0	0.1	0.5
Synchro	0.0	−0.1	0.2	−0.1	0.2	0.5	0.1	0.5
PI	0	−0.2	0.2	−0.1	0.3	0.5	0.1	0.5
Sustainable tower	−0.5	−0.1	0.1	0.1	0.2	0	0.1	0.5
Sustainable synchro	−0.7	−0.5	0.2	−0.1	0.2	0.5	0.1	−0.1
Sustainable PI	−0.7	−0.5	0.2	−0.2	0.2	0.5	0.1	−0.5

Table 12.1. *(Continued)*

Initial Rates of Change for Logistics Operators and Logistics Systems								
Name	GHG	Cost	Traffic	Time	Data	Auto	Resilience	Central
Scenario 2: Initial rates of change for logistics operators								
3PL	−1.0	0.1	0.1	0.0	1.0	1.0	0.0	0.0
Tower	0.0	−0.1	0.1	0.0	0.2	0.0	0.1	0.5
Synchro	0.0	−0.1	0.2	−0.1	0.2	0.5	0.1	0.5
PI	0	−0.2	0.2	−0.1	0.3	0.5	0.1	0.5
Sustainable tower	−0.5	−0.1	0.1	0.1	0.2	0	0.1	0.5
Sustainable synchro	−0.7	−0.5	0.2	−0.1	0.2	0.5	0.1	−0.1
Sustainable PI	−0.7	−0.5	0.2	−0.2	0.2	0.5	0.1	−0.5

RESULTS

Scenario 1: Moderate Acceleration in GHG Policy and Digitalisation

In this scenario, there is a strengthening in climate mitigation policy and digitalisation which causes the 3PL logistics regime to change, but change in this regime is slow. The strengthening in climate mitigation policy is strong enough to create a tipping point where supporters move away from the conventional logistics systems to more sustainable systems. The results are shown in Fig. 12.3. There is competition between sustainable variants of the different digitalised structures, and while logistics towers are adopted for a few years, in the medium term, the advantages of synchromodality in terms of long-term cost savings as well as higher system efficiency enable a low emission, synchromodal logistics takes off. A sustainable form of synchromodality is a dominant solution in this system.

Scenario 2: Far-reaching Change to the Physical Internet

In this scenario, there is both strong policy support for decarbonisation and a more rapid development of digitalisation in a world which quickly realises the benefits of digitalisation in logistics through autonomous vehicles and systems, combined with rapid development of the necessary standards for data exchange and data security. Furthermore, technologies to reduce the large energy demands of blockchain computing technologies are developed. The results are shown in Fig. 12.4.

An interesting outcome is that the rapid development of digitalisation dominates the adoption choices, enabling the development of the physical internet *without* a concentration on GHG emissions reduction. In the longer run, the increasing requirement to reduce emissions enables sustainable synchromodality to take off, promoted by policy to increase the costs of emissions to operators. In the longer term, the physical internet partly replaces the role of synchromodality in decarbonisation with a next step in system-wide rationalisation. Other solutions are pushed to the background with only minor contributions. As in Scenario 1, the role of logistics control towers peaks between 2035 and 2040. Apparently, its importance for decarbonisation depends more on its intrinsic qualities than the presence of other innovations in its environment. Synchromodality peaks well before the final horizon year of 2050, and the physical internet takes the dominant role.

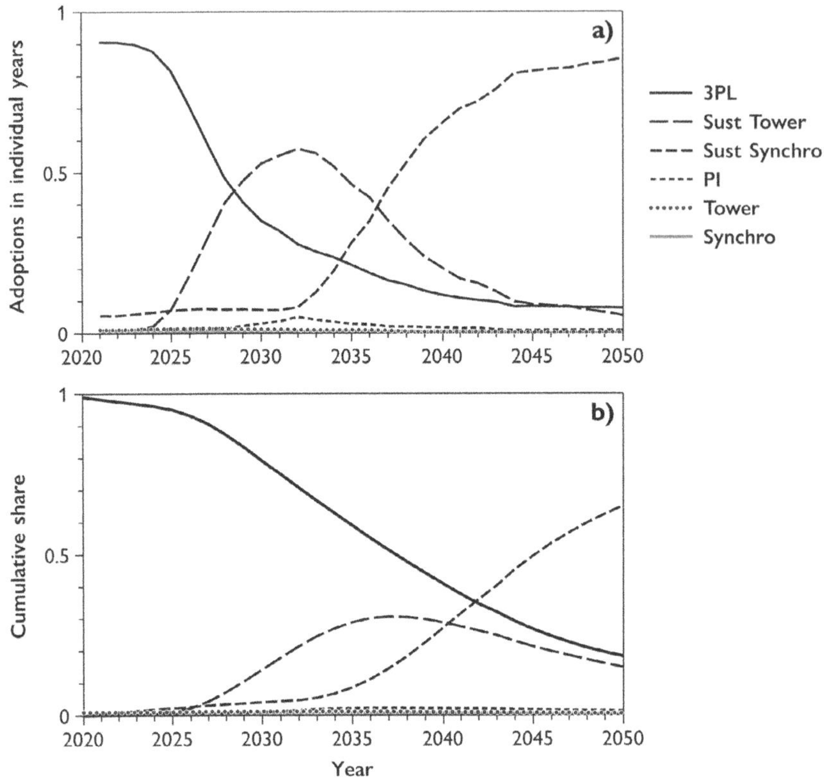

Fig. 12.3. Logistics Companies' Choice of System Types in (a) Individual Years and (b) Cumulatively Over Time, in Scenario 1: Moderate Acceleration in GHG Policy and Digitalisation. *Source*: Authors.

CONCLUSIONS AND FUTURE RESEARCH DIRECTIONS

Our argument in this chapter is that logistics is in a period of fundamental change – a 'twin transition' of digitalisation and decarbonisation. There are important implications for transport policy in terms of climate mitigation policy and the development of the logistics sector as a vital part of the economy. However, the implications of digitalisation for achieving environmental policy goals for climate mitigation have not yet been addressed in policy analysis. Such analysis requires the support of policy modelling, but current transport models have so far not addressed the processes of digitalisation or the interaction between digitalisation and GHG emissions (Tavasszy, 2020).

We propose an approach for how such issues might be addressed in simulation modelling. It uses the concepts of sustainability transitions (Köhler et al., 2019) to derive a theoretical structure for an agent-based simulation model of transitions in logistics, following previous work on sustainability transitions in passenger transport (Köhler et al., 2009). Illustrative scenarios show that in

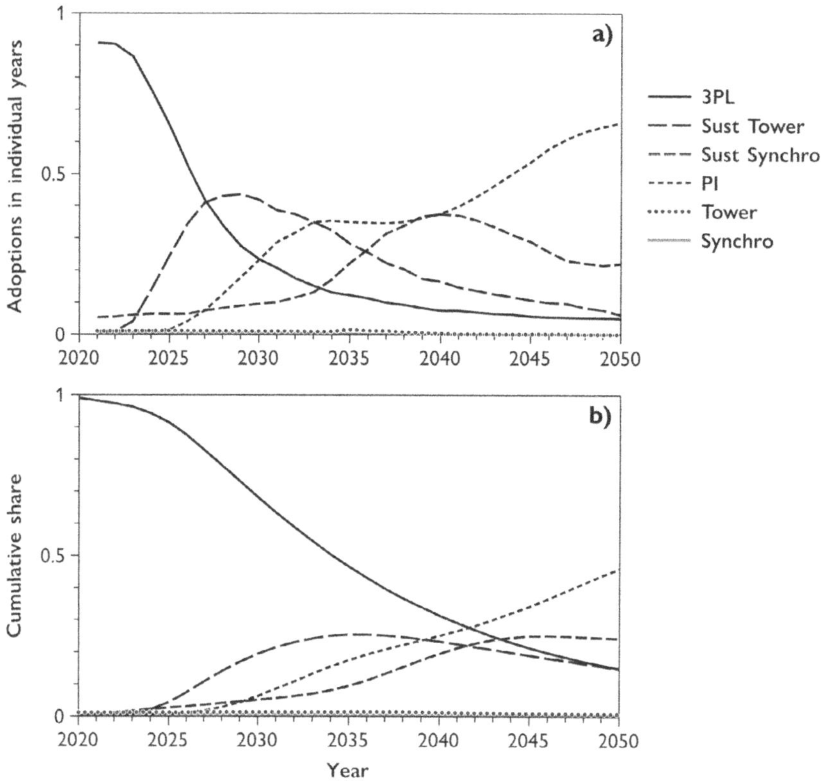

Fig. 12.4. Logistics Companies' Choice of System Types in (a) Individual Years and (b) Cumulatively Over Time, in Scenario 2: Far-reaching Change to the Physical Internet. *Source*: Authors.

comparison to a baseline where change is very slow, transitions to new logistics systems may occur. Current literature and assessments of logistics stakeholders suggest that a 'digitalisation transition' is already under way. The EU emissions trading scheme (ETS) includes transport fuels and therefore strengthens the GHG mitigation policy for the EU, which will require major changes in the EU freight transport system. Therefore, it can be argued that a 'no system change' baseline is not realistic. The transition scenarios illustrated here suggest possible pathways of change, given a current understanding of what future logistics systems could look like.

This analysis is a pilot to show how scenarios of system change in logistics and their results for freight transport can be assessed at an aggregated level. Here, the assessment of future scenarios is based on the literature review and discussions with stakeholders. However, the model could be used in a more comprehensive scenario analysis that combines structured qualitative scenario development with simulation modelling (Köhler et al., 2022).

Tavasszy, L. A. (2020). Predicting the effects of logistics innovations on freight systems: Directions for research. *Transport Policy*, *86*, A1–A6. https://doi.org/10.1016/j.tranpol.2019.11.004

Tavasszy, L. A., Bedhani, B., & Konings, R. (2017). Intermodality and synchromodality. In: H. Geerlings, B. Kuipers, & R. Zuidwijk (Eds.), *Ports and networks - Strategies, operations and perspectives* (pp. 251–266). Routledge.

World Economic Forum. (2016). *World Economic Forum white paper digital transformation of industries: Logistics*. REF 100616. http://reports.weforum.org/digital-transformation/wp-content/blogs.dir/94/mp/files/pages/files/dti-logistics-industry-white-paper.pdf

CHAPTER 13

TOURISM AND TRAVEL: THE CHALLENGES OF NET ZERO

Luca Zamparini

Università del Salento, Italy

ABSTRACT

The international quest for net-zero emissions by mid-century involves tourism activities that, directly and indirectly, contributed about 10% of carbon emissions in pre-COVID-19 years and will probably regress to similar, and possibly higher, values in the future. Despite this, the tourism sector appears to have done very little to prepare itself for the next decades, when accelerating climate change is envisaged. This chapter reviews the main findings with regard to climate change in relevant tourism literature, through an international survey of experiences and deployed strategies to tackle the net-zero emissions challenge in the various world regions and in the various subsectors of the tourism industry. It continues with a comparative evaluation of the various case studies, in order to disentangle similarities and peculiarities in the mitigation strategies implemented, how they have been developed and their degree of effectiveness. Possible future developments at the local, national, and international levels are also discussed.

Keywords: Tourism; net zero; tourism-related transport; tourism supply chains; tourism activities

INTRODUCTION

The international quest for net-zero emissions by mid-century has characterised both theoretical research and administrative activities in the last decades. Peeters and Dubois (2010) showed that overnight tourism represented a considerable share

Towards Transport Net Zero
Transport and Sustainability, Volume 20, 195–208

Published under exclusive licence by Emerald Publishing Limited
ISSN: 2044-9941/doi:10.1108/S2044-994120250000020013

of CO_2 emissions. At the same time, the various possible scenarios of tourism development in the coming decades would not result in emission reductions without radical shifts in policies, strategies, and activities. This has led some to question whether decarbonising tourism would be achievable (Becken, 2019), given various important obstacles: e.g., the growth of the industry, institutionalisation of interests, flawed policies, and the geography of tourism. It is now estimated that tourism activities, directly and indirectly, contributed about 10% of global carbon emissions in pre-COVID-19 years, and they will probably return to similar, and possibly higher, values in the future (January–July 2024 international tourist arrivals were already back to 96% of pre-pandemic levels – see UN Tourism, 2024). Tourism-related transport alone is predicted to account for more than 5% of all global CO_2 emissions by 2030 (UN World Tourism Organization, 2019). At the international level, the Glasgow Declaration of 2021 for *A commitment to a decade of tourism climate action* is noteworthy as it aligns the sector with global commitments and proposes collaborative solutions. Moreover, the International Energy Agency (IEA) has issued a detailed roadmap to a 2050 net-zero future (Scott & Gössling, 2022a) that includes net-zero tourism as one of the important requirements to avoid a global temperature increase of 3°C. In order to reach this goal, all subsectors of the tourism industry will need to be involved, with a particular emphasis on international air travel and tourism growth projections.

These general principles notwithstanding, the tourism sector appears to have done very little to prepare itself for the next decades, when accelerating climate change is envisaged. This also applies to the last decades of research according to Scott and Gössling (2022b). Moreover, tourism is vulnerable to climate change in several respects (Jones, 2023): rising temperatures will jeopardise the season of winter sports, reductions in biodiversity will hamper ecotourism activities, rising sea levels can threaten coastal tourism infrastructures and resorts (Khanal et al., 2022), and so on. Yet the development of an environmentally friendly tourism system can also determine a competitive advantage in the world tourism market. The importance of fostering ecotourism activities for a sustainable and net-zero evolution of this industry has been examined by Kwok (2022), who proposes a strategy based on six main pillars: (a) reduction of energy demand through performance-based estimations of clients' expectations, occupancy, and intended use; (b) reduction of embodied carbon in the planning and construction phases of buildings; (c) low carbon energy and renewable energy supply via the use of on-site generation of heat and water, and efficient storage systems; (d) zero-carbon policies at all administrative levels; (e) environmental assessment tools and ecotourism certifications; and (f) monitoring activities for the assessment of energy consumption. The taxonomy is worth mentioning because of its comprehensiveness.

A study by Jones (2023) highlighted that many tourism companies are trying to address the transition towards net-zero emissions, although with varying levels of commitment among firms and subsectors of tourism. Considering that, after the COVID-19 period, another structural growth trend is forecast, cooperation, and cross-agency collaboration to design and deploy consistent initiatives that promote systemic change to allow decarbonisation will be essential (Higham et al., 2022). Transdisciplinary partnerships need to be fostered to enhance climate action and to develop policies consistent with climate commitments, to make better use of local knowledge, and to increase scale information to guide climate plans (Scott, 2021). According to Gössling et al. (2023), it is necessary to consider four dimensions – scales, scopes, stakeholders, and strategies – in order to deploy mitigation strategies. To explain Gössling's terms, scale implies the need to consider all levels (global, national, destination, business) at which such strategies can be implemented; scope identifies the emissions to be considered and those to be excluded; stakeholder refers to the accountability of all those agents that may be in charge of reducing emissions; and strategies account for the mechanisms of emissions reduction.

Considering that the tourism market is constituted by a series of complementary goods and services, it is important to disentangle the various subsectors of tourism, their interrelationships, and their contribution to climate change. Consequently, it is necessary to identify relevant stakeholders within the tourism industry and to assess how they are addressing issues of decarbonisation. This chapter aims to provide a review of the main findings of relevant tourism literature by undertaking an international survey of experiences and deployed strategies to tackle the net-zero-emission challenge in the various world regions and in the various subsectors of the tourism industry. Consequently, the following sections will be devoted to the analysis and discussion of the role of (a) transport, (b) accommodation facilities, (c) catering facilities, and (d) experiences and activities, in reaching the goal of net-zero emissions in tourism. The last section of this chapter offers a comparative evaluation of the discussions presented in the previous sections, and it also proposes possible future developments at the local, national, and international levels.

THE ROLE OF TRANSPORT FOR NET-ZERO EMISSIONS IN TOURISM

The important role of transport choices and policies for the achievement of net-zero emissions in tourism clearly emerges if the CO_2 emissions per person per km of the various transport modes are considered (Fig. 13.1). Buses, trains, and minibuses are the most sustainable modes, with values clearly lower than all other (private) motorised means of road transport (car, motorcycle, campervan, and jeep), aeroplanes, and cruise ships. In the case of planes, the impact is inversely correlated to the travelled distance. The contribution of short-haul flights ($<$500 km) to CO_2 emissions is 10 times higher than buses and 7 times higher than trains.

Eijgelaar and Peeters (2018) sketch some of the main themes that are significant in reducing transport carbon emissions in the tourism sector: (a) the presence of charging infrastructures at tourism destinations; (b) the perception by tourists of the various possible transport modes both to reach the destination and within it; (c) the importance of ICTs and online travel products to foster the choice of more sustainable transport options; (d) the promotion of rail travel through easier ticketing, interoperability, and high-speed rail investments; and (e) heterogeneous tax regimes among transport modes on the basis of their impact

through operational and technological means, and favourable aviation policies should all be developed to increase the sustainability of aviation. Moreover, strict coordination among countries is another necessary requirement, alongside cost analysis and the optimisation of the fleet composition of airlines.

THE ROLE OF ACCOMMODATION FACILITIES FOR NET-ZERO EMISSIONS IN TOURISM

The pledge to upgrade buildings in order to reduce their carbon footprint and, ideally, to bring them to net-zero emissions is common, in the first decades of the 21st century, to most of the world's regions. This theme has, consequently, involved hotel and other types of accommodation facilities in tourism. In mature destinations, this is a particularly important issue as many structures were built several decades before the control and reduction of carbon emissions were important considerations. Causone et al. (2021) have proposed a survey of the definitions of carbon-neutral and zero-carbon buildings and provide the following definition: a carbon-neutral building is defined as a building able to achieve a net amount of carbon dioxide equivalent emissions related to the whole building life cycle equal to zero.

One of the first pertinent studies in tourism has been related to the Balearic Islands (Rossellò-Batle et al., 2010). Most of the hotels that were present in this mature tourism destination were built in the 1970s–1980s and were, consequently, of low energy efficiency standards compared with more modern equivalents. The authors suggest that the best possible upgrade options would be solar, biomass, or trigeneration systems, but the high investment costs to install such equipment make it very difficult for small and medium hotels. Rossellò-Batle et al. propose a series of actions to promote the move towards lower carbon footprints and, ideally, net-zero energy consumption. First, the environmental benefits related to the reductions of CO_2 emissions thanks to solar and biomass technologies should be economically estimated and internalised through fiscal incentives. In addition, they suggest the excess production of renewable energy could be transferred to the grids of local destinations and economically compensated in order to reduce the payback period of the investments.

Pablo-Romero et al. (2019) analysed the tourism and temperature effects on the electricity consumption of the hospitality sector by considering a sample of hotels in 12 Spanish Mediterranean provinces. It has emerged that there is a strong positive correlation between overnight stays and electricity consumption, especially in high-rated hotels that host foreign and high-income tourists who are not impacted by the energy price. As expected, there is also a positive correlation between temperature and energy consumption. The authors suggest measures to reduce consumption and the transition to renewable energy sources should be implemented, among them so-called deseasonalisation policies on the basis that in milder climate seasons, there is a reduction in energy consumption. Another European case study considers a hotel in Milan, Italy. Causone et al. (2021) state that the efficiency in operational energy consumption is not sufficient and that there should be an active selection of materials that provide the best performance in the entire life cycle. Moreover, in common with other studies, they assume that the extra costs in the building/refurbishment phase should be offset in the operational phase of the hotel.

On the other hand, Vourdoubas (2018) has questioned the feasibility of net-zero-carbon emissions in hotels in the Mediterranean region. The analysis in the paper confirms that the use of renewable energy sources in Mediterranean hotels is rather limited even though this region has abundant resources. Vourdoubas argues large numbers of hotels should be connected to grids that are based on renewable energy resources that would provide all necessary energy, and that this will probably be more easily accomplished if marketing is targeted at environmentally conscious tourists who are eager to offset their carbon footprint and/or willing to pay higher prices for sustainable accommodation facilities. The barriers that make this transition difficult are related to the lack of awareness among hotel managers and owners and the low involvement of public administration in various regions and countries.

Some studies have analysed the possibility of carbon footprint reduction by African hotels. Diab et al. (2015) examined a case study of a hypothetical environmentally friendly tourist village in Alexandria, Egypt, based on a hybrid renewable energy system that would host up to 1,000 guests. The hybrid system would be constituted by photovoltaic (PV) panels, wind turbines, a diesel generator, power converters, and batteries. The authors conclude that the hybrid system is the optimal solution according to a series of parameters such as economic cost, net present cost, and greenhouse gases emitted. Abdelhady's (2023) study also relates to Egypt and considers the optimal hybrid renewable energy system design for a hotel building aiming to reach net zero. This time the system would include PV panels, wind turbines, and biogas generators connected to a grid that optimised the net present cost of the investment and reduced the price of energy. The analysis implies that in such an environment the biogas generators have the lowest level of economic efficiency, given that the equipment has high operating costs. Moreover, in order to be economically feasible, the system would have to generate an excess of energy that could be sold to the local destination grid.

Oluseyi et al. (2016) considered energy consumption trends in Nigeria and assessed the carbon footprint of the hotel sector in the capital city of Lagos. Their analysis ascertained that the accommodation sector in Nigeria is marked by high levels of CO_2 emissions. The most important explanatory variables are total floor area, number of guest rooms, and number of employees, while age and occupancy rate seem to have a lower impact. Consequently, it is necessary to adopt emission reduction policies, and Oluyesi et al. contend this is best achieved by the public administration and hotel sector stakeholders working in partnership. As part of their study, the authors also provide a comparative analysis of the energy use index (KWh/m^2/year) of hotels in various countries (Fig. 13.2). As the figure shows, Hong Kong was the country with the highest energy use among those surveyed in repeated studies, followed by Turkey, Italy, and South Korea. Greece, Vietnam, and the Hainan province in China have the least energy-consuming hotels.

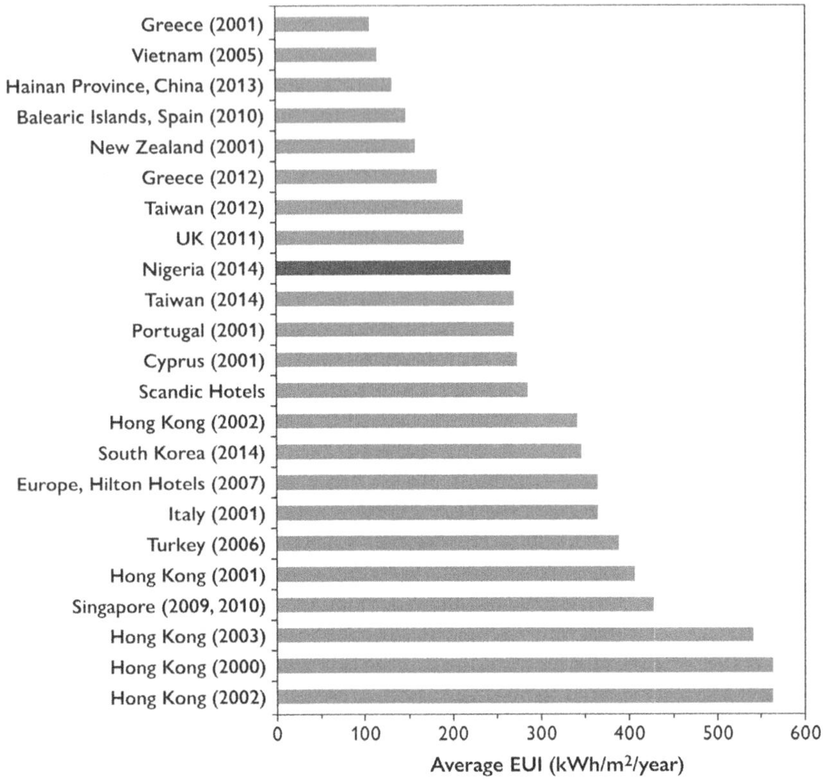

Fig. 13.2. Comparison of Energy Use Index in Various Countries. *Source*: Oluseyi et al. (2016).

THE ROLE OF CATERING FACILITIES FOR NET-ZERO EMISSIONS IN TOURISM

Another significant element of the tourist experience is represented by food and eating activities. The way in which eating parlours are built, and their operational activities, can have an appreciable impact on the carbon footprint of tourist destinations. In recent decades, several studies have evaluated this topic in a number of different settings. Conner et al. (2014) focused on a proposed facility in Chicago, USA, for a market leading quick service restaurant chain. They argued that it would be possible to devise a net-zero energy building, and that the necessary design steps would be the identification of building constraints, the maximum renewable electricity generation potential, the determination of the renewable sources to meet non-electricity energy loads, the identification of the most updated technologies, the softening of constraints, and the final implementation. They conclude that it is important to choose optimal material in the construction of the facility, to adopt energy saving lighting equipment, and a geothermal heating and cooling system. The adoption of all these infrastructures would reduce energy consumption by about 50%. What is lacking in this study is the appraisal of the waste management system, which constitutes an important part of the environmental sustainability of restaurants.

Hai et al. (2023) undertook a similar study where they pondered the possibility of modelling an energy system for a near-zero energy restaurant with green hydrogen energy storage systems in Doha (Qatar). They assumed that energy would be produced by using solar panels and electricity grids. Their study also evaluated the thermal comfort of customers inside the facility and simulated its refrigerating infrastructure, given weather data and information about the restaurant. Hai et al. concluded that the best possible option includes 71% of energy provided by solar panels, with the remaining 29% by the electricity grid. They also state that further optimisation of the energy use in a restaurant may involve the use of electricity for cooking (this would avoid gas) and the recovery of lost heat.

Lund Christiansen et al. (2023) propose a similar analysis related to a Swedish fast food parlour chain. They state that, normally, companies' commitments to the reduction of carbon emissions rely mainly on offsetting activities that lead to business-as-usual practices rather than getting involved in more efficient opportunities to minimise the carbon footprint. It also appears that the voluntary and individualised approach does not seem to lead to significant changes in operations and strategies. There emerges the need for regulations that should establish targets for both carbon removals and energy offsetting, because otherwise projects for carbon offsetting only compensate for emissions that could be eliminated. This would nudge companies to prioritise options that reduce emissions instead of searching only for cost-efficient ones.

Kuuluvalainen et al. (2022) have carried out a study that aims at analysing all themes related to the achievement of a net-zero-carbon emission restaurant. They state that there are several reasons to reduce the carbon footprint: risk management, money savings, and the achievement of a responsible and attractive image as an employer towards employees. Among the strategies to reach net zero is the choice of ingredients (animal origin foods should be replaced by plant-based ones) and the avoidance of frozen foods and ingredients from remote countries. The net-zero restaurant would also optimise its activities in order to reduce food waste as much as possible. Moreover, given that the energy consumption of restaurants is about five times higher than the average of other business facilities, making it more efficient implies significant cost savings. The study also proposes some marketing strategies that may be adopted by restaurants to attract customers, emphasising the role of digital solutions, on the basis that in general environmentally friendly procedures and strategies may give a competitive advantage to restaurants pursuing them.

Along the same lines is a study by Tehrani et al. (2020) that considers the role of the restaurant industry within green cities and waste management. They also assume that restaurant owners may consider the positive impact of sustainable activities and strategies in attracting customers. Their research focuses on the possible use of waste materials as generator of energy. On the basis of a structured sample of eating facilities in two US cities (Providence and Springfield) and

tested. The need for sound methodologies to estimate the impact of supposed net-zero activities in all the dimensions of tourism is evident, and several authors have provided possible analytical tools.

Many contributions have also highlighted the coexistence of different stakeholders that may claim heterogeneous priorities and preferences, and the use of multi-criteria decision analysis may be very effective in providing an assessment of the mitigation and adaptation options (Michailidou et al., 2016). Such methodology may be very efficient in contexts where it is necessary to identify and define the main assessment criteria, to apply to each of them an appropriate weighting factor, to fix the acceptable thresholds for sensible environmental parameters, and finally to rank the possible alternatives in order to choose the one that may be most efficient to reduce the carbon footprint of an investment/activity in tourism. Multi-criteria analysis can also be adapted to specific regional or contextual situations, by taking into account all of their peculiarities in the quest of the optimal solution.

Ultimately, a shift of paradigm appears as the most important element to try to decarbonise the tourism industry. Technological developments alone are not sufficient. Changes at the bottom by consumers, entrepreneurs, or innovators may represent further important elements to devise more sustainable solutions. In this context, the role of training of employees and of employers constitutes an important element in order to raise awareness and to promote sustainable strategies and procedures. Both incremental and the radical innovations can then be institutionalised by the relevant administrative bodies and diffused and adapted to various specific contexts. Moreover, a holistic approach should encompass all the phases and markets of the tourism activities in a synergistic and proactive way.

REFERENCES

Abdelhady, S. (2023). Techno-economic study and the optimal hybrid renewable energy system design for a hotel building with net zero energy and net zero carbon emissions. *Energy Conversion and Management*, *289*, 117195.

Asghari, M., Mirzapour Al-e-hashem, S., & Mohammad, J. (2021). Green vehicle routing problem: A state-of-the-art review. *International Journal of Production Economics*, *231*, 107899.

Becken, S. (2019). Decarbonising tourism: mission impossible? *Tourism Recreation Research*, *44*(4), 419–433.

Betta, L., Dattilo, B., di Bella, E., Finocchiaro, G., & Iaccarino, S. (2021). Tourism and road transport emissions in Italy. *Sustainability*, *13*, 12712.

Blanco, E. E., & Sheffi, Y. (2017). Green logistics. In Y. Bouchery, C. Corbett, J. Fransoo, & T. Tan (Eds.), *Sustainable supply chains* (Springer Series in Supply Chain Management, Vol. 4, pp. 147–187). Springer.

Causone, F., Tatti, A., & Alongi, A. (2021). From nearly zero energy to carbon neutral: Case study of a hospitality building. *Applied Sciences*, *11*, 10148.

Conner, E., Ramirez Milan, M., & Wallace, L. (2014). *Net zero energy building analysis for McDonalds USA, LLC*. Duke University.

Diab, F., Lan, H., Zhang, L., & Ali, S. (2015). An environmentally-friendly tourist village in Egypt based on a hybrid renewable energy system – Part two: A net zero energy tourist village. *Energies*, *8*, 6945–6961.

Eijgelaar, E., & Peeters, P. (2018). *Zero emission tourism mobility. A research and policy agenda*. Centre for Sustainability, Tourism and Transport, Breda University.

Garvey, A., Norman, J. B., Owen, A., & Barrett, J. (2021). Towards net zero nutrition: The contribution of demand-side change to mitigating UK food emissions. *Journal of Cleaner Production, 290*, 125672.

Gössling, S., Balas, M., Mayer, M., & Sun, Y. Y. (2023). A review of tourism and climate change mitigation: The scales, scopes, stakeholders and strategies of carbon management. *Tourism Management, 95*, 104681.

Gössling, S., & Humpe, A. (2023). Net-zero aviation: Time for a new business model? *Journal of Air Transport Management, 107*, 102353.

Guhnemann, A., Kurzweil, A., & Mailer, M. (2021). Tourism mobility and climate change –A review of the situation in Austria. *Journal of Outdoor Recreation and Tourism, 34*, 100382.

Hai, T., Alenizi, F. A., Hussein Mohammed, A., Kumar Singh, P., Metwally, A. S. M., & Ullah, M. (2023). Optimal design modeling of an energy system for a near-zero energy restaurant with green hydrogen energy storage systems. *Fuel, 351*, 128896.

He, Y., Huang, P., & Xu, H. (2018). Simulation of a dynamical ecotourism system with low carbon activity: A case study from western China. *Journal of Environmental Management, 206*, 1243–1252.

Heggen Johansen, B. (2021). Establishing a basis for decision support modelling of future zero emissions sea based tourism mobility in the Geiranger fjord area. *ECMS, 35*(1), 1–7.

Higgins-Desbiolles, F. (2023). Subsidiarity in tourism and travel circuits in the face of climate crisis. *Current Issues in Tourism, 26*(19), 3091–3101.

Higham, J., Font, X., & Wu, J. (2022). Code red for sustainable tourism. *Journal of Sustainable Tourism, 30*(1), 1–13.

Jones, P. (2023). Towards net-zero emissions in the global tourism industry. *Athens Journal of Tourism, 10*(2), 125–136.

Kamb, A., Lundberg, E., Larsson, J., & Nilsson, J. (2021). Potentials for reducing climate impact from tourism transport behavior. *Journal of Sustainable Tourism, 29*(8), 1365–1382.

Khanal, A., Mafizur Rahman, M., Khanam, R., & Velayutham, E. (2022). Does tourism contribute towards zero-carbon in Australia? Evidence from ARDL modelling approach. *Energy Strategy Reviews, 43*, 100907.

Kuuluvalainen, S., Paallysaho, M., Rajala, T., Risu, E., Turunen, A., & Vitikka, M. (2022). *Restaurants and carbon footprint mission zero foodprint workbook*. Laurea University of Applied Sciences.

Kwok, T. K. (2022). Concept of net zero ecotourism in sustainable tourism industry development. *International Journal of Social and Business Sciences, 16*(8), 413–416.

Liu, Y., Jiang, Q., & Gleasure, R. (2023). Hitting net-zero without stopping flying: Increasing air travelers' likelihood to opt-in to voluntary carbon offsetting. *Journal of Travel Research, 62*(1), 21–38.

Lund Christiansen, K., Hadju, F., Planting Mollaoglou, E., Andrews, A., Carton, W., & Fischer, K. (2023). "Our burgers eat carbon": Investigating the discourses of corporate net-zero commitments. *Environmental Science and Policy, 142*, 79–88.

Michailidou, A. V., Vlachokostas, C., & Moussipoulos, N. (2016). Interactions between climate change and the tourism sector: Multiple criteria decision analysis to assess mitigation and adaptation options in tourism areas. *Tourism Management, 55*, 1–12.

Mulholland, E., Teter, J., Cazzola, P., McDonald, Z., & Gallachóir, B. P. Ó. (2018). The long haul towards decarbonising road freight – A global assessment to 2050. *Applied Energy, 216*, 678–693.

Müller, A. (2023). Decarbonizing business travel: a qualitative exploration of the (mis)alignment between knowledge organizations' climate strategies and travel practices. *Journal of Sustainable Tourism, 32*(8), 1557–1583.

Nhamo, G., Dube, K., Chapungu, L., & Chikodzi, D. (2023). Quest for netzero emissions in South African national parks: A tourism perspective. *Heliyon, 9*, e16410.

Oluseyi, P. O., Babatunde, O. M., & Babatunde, O. A. (2016). Assessment of energy consumption and carbon footprint from the hotel sector within Lagos, Nigeria. *Energy and Buildings, 118*, 106–113.

Pablo-Romero, M., Pozo-Barajas, R., & Sanchez-Rivas, S. (2019). Tourism and temperature effects on the electricity consumption of the hospitality sector. *Journal of Cleaner Production, 240*, 118168.

Peeters, P., & Dubois, G. (2010). Tourism travel under climate change mitigation constraints. *Journal of Transport Geography, 18*, 447–457.

Peeters, P., Higham, J., Cohen, S., Eijgelaar, E., & Gössling, S. (2019). Desirable tourism transport futures. *Journal of Sustainable Tourism*, *27*(2), 173–188.

Probstl-Haider, U., Mostegl, N., & Damm, A. (2021). Tourism and climate change – A discussion of suitable strategies for Austria. *Journal of Outdoor Recreation and Tourism*, *34*, 100394.

Rossellò-Batle, B., Moia, A., Cladera, A., & Martinez, V. (2010). Energy use, CO_2 emissions and waste throughout the life cycle of a sample of hotels in the Balearic Islands. *Energy and Buildings*, *42*(4), 547–558.

Ruhlin, V., & Scherrer, M. (2023). Towards net zero emissions logistics cold chains – An early stage assessment of GHG reduction potentials in the fruits and vegetables industry. *Transportation Research Procedia*, *72*, 1105–1112.

Scott, D. (2021). Sustainable tourism and the grand challenge of climate change. *Sustainability*, *13*, 1966.

Scott, D., & Gössling, S. (2022a). Destination net-zero: What does the international energy agency roadmap mean for tourism? *Journal of Sustainable Tourism*, *30*(1), 14–31.

Scott, D., & Gössling, S. (2022b). A review of research into tourism and climate change – Launching the annals of tourism research curated collection on tourism and climate change. *Annals of Tourism Research*, *95*, 103409.

Tehrani, M., Fulton, L., & Schmutz, B. (2020). Green cities and waste management: The restaurant industry. *Sustainability*, *12*, 5964.

UN Tourism. (2024). *UN Tourism Barometer*. UN Tourism World Tourism Barometer|Global Tourism Statistics.

UN World Tourism Organization. (2019). *Transport related CO_2 emissions of the tourism sector*. Retrieved January 17, 2025, from https://www.e-unwto.org/doi/book/10.18111/9789284416660

Vourdoubas, J. (2018). Hotels with net zero carbon emissions in the Mediterranean region: Are they feasible? *Journal of Tourism and Hospitality Management*, *6*(2), 72–79.

Wang, H., Sheng, N., Song, Q., Xu, L., & Bai, J. (2024). Exploring the potential carbon emissions and net-zero path of international flights: A case study in Macao. *Journal of Environmental Sciences*, *156*, 288–299.

CHAPTER 14

URBAN TRANSPORT IN NET ZERO CITIES

Maria Attard

University of Malta, Malta

ABSTRACT

Urban areas are home to over half of the world's population. Cities and their transport systems continue challenging decarbonisation; however, there are also other impacts from increased levels of noise pollution to road accidents, transport poverty, and social exclusion. This chapter seeks to address a number of themes and transport modes which are considered important for the future of urban transport. The role of public transport, active mobility, and micromobility is discussed as critical infrastructures that support a just and net zero city. The increasing movement of goods within cities requires more attention, primarily because of new consumption patterns, the rise in e-commerce, and small urban deliveries. The promise of multimodality is discussed, as well as the link between land use and transport planning. This chapter also discusses the future smart city and the role of technology in delivering necessary solutions. It concludes with some thoughts about the future urban transport system and the need to re-focus planning on human-scale cities that deliver the transition to net zero in a safe and equitable way.

Keywords: Urban transport; public transport; urban freight transport; walking; micromobility; sustainable urban mobility plans

Towards Transport Net Zero
Transport and Sustainability, Volume 20, 209–223

Published under exclusive licence by Emerald Publishing Limited
ISSN: 2044-9941/doi:10.1108/S2044-994120250000020014

Despite this, infrastructure to support behaviour change remains a necessity. While in many cities, especially those in the Global North, there are relatively well-established systems for public transport, the differences in the provision and quality of services are significant. The *Union Internationale des Transports Publics* (UITP, 2017) identifies the bus as being the dominant mode of public transport in many countries in the world, serving 63% of all public transport journeys in 2015, compared to metros (16%), suburban rail (16%), and trams (5%). The focus on the bus is primarily due to its flexibility but also its cost. In a growing number of cities, the implementation of bus rapid transit (BRT) has highlighted even further the important role of the bus in providing safe, affordable, and accessible public transport services. With over 190 cities having implemented BRT systems across all regions of the world (see www.brtdata.org), the potential for such public transport infrastructure can be further enhanced through technologies such as electrification and automation.

While the status and role of public transport infrastructure in cities are relatively understood within broader transport and land-use policies, and in regulations governing public transport operations and contracting, the role of the complementary modes of walking, cycling, and micromobility remains more elusive (see Chapters 3 and 4). Over recent years, policy discourse has emphasised more rigorously the importance of active travel, highlighted by the COVID-19 years and the increase of demand for walking infrastructure in lieu of travel restrictions. Needless to say, these modes of transport are a key solution to decarbonisation in urban areas and should be the primary mode of transport for short trips in any city or as a complementary last-mile mode of travel for public transport systems. However, the gap between the scientific evidence and the policy imperatives of supra-national organisations (including the World Health Organization, the EU, and the UN) that keep insisting on the need to prioritise walking and cycling to combat rising emissions, and the action on the ground in many cities across the Global North and the Global South is substantial. Chapter 19 in this volume describes the efforts on active travel during the pandemic, from Open Streets and Slow Streets programmes in the United States, to measures implemented in Europe and over 900 interventions recorded across 400 cities (Combs, 2024), but the difficulties of active travel infrastructure provision in cities remains challenging. Research into urban space distribution further supplemented the discussion by highlighting the evident focus on many of our urban transport systems to cater primarily for the car (Attard et al., 2023; Gössling, 2016; Gössling et al., 2016; Nello-Deakin, 2019; Urry, 2013).

On the other hand, the verdict is still out on the future of some micromobility modes in cities. The uneven and at times contested deployment and operation of some micromobility services has left many wondering what future holds for these modes. Over the past decade, there has been a relatively fast deployment and adoption of e-bicycles and e-scooters, both offered as part of shared services by various platform operators or even privately owned. The availability of shared e-scooter services has spread across numerous cities, and their impact has been the subject of many studies since they hold a potential to support public transport by providing a service where buses do not reach but also to expedite travel where walking might take longer or (car) traffic is too congested (Aarhaug, Fearnley, Hartveit, et al., 2023; Aarhaug et al., 2023; Lu et al., 2021). Some authors have identified concerns related to a modal shift from walking and public transport use, rather than car, and this has negative impacts on public transport demand as well as overall emissions (Aarhaug et al., 2023; Attard & Balbontin, 2024).

In the more effective cases, e-scooter introduction was followed up by new regulatory frameworks and changes, as services flourished and increased. The work of Aarhaug, Fearnley, Hartveit, et al. (2023) shows how in Norway e-scooters were introduced in a liberal regulatory environment, classifying them as bicycles and limiting the speed to 20 kph. Subsequent regulatory interventions supported the growth of these services. In other contexts, piloting and monitoring has ensured a more managed approach such as the case of UK cities (ARUP & NatCen, 2022). At the same time, other cities have banned shared e-scooters, Paris being the most high-profile case, and moved on to promote privately owned e-scooter and e-bicycle use through subsidy schemes for battery-operated personal micromobility vehicles (e-scooters, e-bicycles, and e-motor scooters), as is the case in Malta (D'Orso et al., 2024).

MOVEMENT OF FREIGHT IN URBAN TRANSPORT

The global city population doubled from 25% in 1950 to over 50% in 2020 (United Nations Human Settlements Programme, 2022). Production, consumption, and human activity continue to concentrate and grow further around cities, and a consequence of this is that urban freight systems are becoming increasingly evident in their contribution to transport impacts in cities. Interest in the field of sustainable logistics has increased over the years (Chapter 12), but actions on the ground have been slower.

Furthermore, following the COVID-19 pandemic, there has been a rise in urban deliveries that continue to challenge the mitigation of emissions from transport. Urban delivery trips are generally short and frequent and carry small loads to increase distribution and speed. Changing consumer patterns and increasing last-mile deliveries (including home deliveries) occur without operational efficiency. Indeed, many home deliveries are repeated because of failed delivery or returns with much waste through the supply chain (Wygonik & Goodchild, 2018). Urban delivery trips represent only 3% of the total freight activity but account for 20% of emissions from the sector (Chakravarty, 2014). Depending on the origin and destination of goods, urban freight occurs at both a local and regional scale and uses urban space for the movement between terminals, distribution hubs, and customer locations. The rise of e-commerce has literally reallocated the movement of goods from retail centres to individual homes and residential areas, and with the EU's goal of achieving net zero city logistics by 2030 (European Commission, 2013), the challenges are many (Attard, 2023).

Bjørgen et al. (2019) identify several challenges to coordinate logistics planning at the local level. Indeed, many local authorities continue to ignore freight movement in urban mobility plans. There is a realisation that the adoption of new technologies alone (e.g., electrification of the fleet) will not be enough because of embedded carbon impacts and slow transitions, and therefore policies, regulation and practices must change at the same time (Geels, 2012). There is an increased attention at the EU level of the requirement to manage effectively the movement of goods, not only as a requirement to decarbonise but also as part of the EU's urban mobility package and the integration of intelligent transport systems for optimising deliveries and digitising the entire freight system (Aifandopoulou & Xenou, 2019). The *Sustainable Urban Logistics Planning* Topic Guide was published in 2019 as part of the Horizon Project NOVELOG (https://civitas.eu/projects/novelog) and presents a toolkit, similar to the first SUMP guidelines of 2013, for cities to develop sustainable urban logistics plans (SULPs). Adopting a similar approach to the SUMP, the SULP guidance identifies the complex nature of urban freight transport and provides a methodology that includes (i) the analysis of the status quo, (ii) the development of a strategy through stakeholder engagement, (iii) the planning frameworks required for successful adoption, and (iv) approaches to implementation and monitoring.

The impact of innovation in the freight sector is highlighted by several studies that show, among other things, that electrification and new technologies will enable heavy trucks to be more efficient. In their net zero scenario, the International Energy Agency (IEA, 2021) estimated the sales of battery-electric and fuel-cell trucks to increase above 50% by 2030 and the efficiency of conventional trucks to improve. Other technologies such as drone delivery, using uncrewed aircraft that carry and transport parcels in specified locations (Wang et al., 2022), have the potential to reduce the cost of the last mile, which is identified as the most expensive phase in the supply chain. McKinsey (2023) reported that a single parcel delivery cost would be significantly less than even an electric car if drones are coordinated and managed simultaneously. There are fuel savings, even compared to electric vehicles, as well as additional gains in areas with poor transport infrastructure or congested urban roadways.[2]

Uptake of logistics technologies, however, is not without challenge. Studies have shown issues with customer acceptance (Cai et al., 2021), as well as concerns about human interaction (Stephan et al., 2022). Understanding the underlying behavioural barriers to the deployment of such technologies assists their faster adoption. In some instances, however, adoption needs to be expedited, encouraged, enforced even, so that technological benefits are not lost in time.

THE PROMISE OF MULTIMODALITY

Because cities condense human activities within a specific area, they provide several opportunities for coordinating existing (and future) transport services into one integrated multimodal offer. Multimodal transport links together mobility solutions, enabling commuters and travellers to combine a variety of modes to reach a destination. The integration of the bus with the train or water taxis and sea ferries or even use of flexible transport services to navigate the distance between journey origin and destination points offers many benefits, especially for people living in dense but spatially diverse cities such as London, Paris, and Milan in Europe, or Beijing and Shanghai in China. Intra-city travel could be facilitated by coordinating trains with local bus or metro services (e.g., Channel Tunnel in London and Paris). Mobility-as-a-service (MaaS) and shared mobility platforms like Bolt (www.bolt.eu), Uber (www.uber.com), and Moovit (www.moovit.com) integrate payment and offer advantageous alternatives to private car use, especially in congested and expensive city centres. The shift to such services has the potential to contribute significantly to net zero, future cities, especially when combined with electric, autonomous, and shared systems (Ding & Jian, 2024).

The promise of multimodality is, however, challenged by a number of issues. First, and probably the most important, is the necessity for an overarching framework that allows seamless multimodal journeys for the users. Second is that if the services are offered by various service providers, collaboration between operators is needed. There is fierce competition between public transport operators which now operate at a global level and compete for city contracts across different geographies, MaaS operators that aim to dominate city markets with their service offers, and local operators that offer niche services, for example, traditional water taxis or gondola-style transport where cities benefit from access to the sea or river water. This is the case in the islands of Malta where the protected and historically rich Grand Harbour offers ample opportunity for increasing connectivity and access to the capital city Valletta. Integration between sea ferries operations and public transport services is necessary if multimodality is to be achieved.

Tourist destinations are particularly well suited for such multimodality as they make it possible to combine traditional and newer services for thousands, if not millions, of tourists searching for cheap but comfortable means of visiting a city. The ubiquitous adoption of smartphones alongside the availability of WiFi has pushed for the development of smartphone applications that provide information and trip planning services for tourists but also for locals (Maas et al., 2021). The power of information to influence behaviour towards more environmentally sustainable modes of transport has been studied and forms part of the increasing body of literature that looks at using ‘nudges’ to influence behaviour (Attard, Cañas et al., 2024). These measures, however, are linked to other city systems and are especially sensitive to land use and development planning.

Finally, there is also some hope for integration of modes that could facilitate freight movements. Several authors have investigated the possibilities of passenger and freight transport integration (Antoniou et al., 2023; Bruzzone, 2021; Cavallaro & Nocera, 2022). The need for various small-scale services and dialogue between stakeholders is undoubtedly necessary to address operational but also corporate issues, users’ needs, and policy requirements. Antoniou et al. (2023) identify the need for further analysis into the contribution of integrated passenger and freight transport in view of environmental and social sustainability goals.

car, and 27% likening the concept to an automated freight truck. As detailed by Thomopoulos et al. (2024), many authors have provided several definitions, some more specific to projects and initiatives and others which are issued and dictated by standards (e.g., SAE J3016 and ISO, 2021). Indeed, Hopkins and Schwanen (2021) argue that the current discourse on autonomous vehicles is too techno-centric and fails to account for the nuances affecting user experiences and human/non-human interactions, raising concerns about the potential lock-in effects produced by technologies that exclude rather than include other solutions. Indeed, factoring in basic needs and limiting the impacts on pollution and traffic are probably the two main concerns voiced by several researchers with respect to ensuring the sustainable deployment of autonomous and connected vehicles (Attard, Thomopoulos et al., 2024; Beyazit et al., 2024; Nikitas, 2024). The issues of equity and access, and an implementation that contributes to building resilient cities and ensuring good governance and regulation, are all critical in the deployment of autonomous vehicles.

Concerns have also been raised about the aspirations linked to the electrification of vehicle fleets, in an attempt to curb tailpipe emissions. EU policy and funding have been substantial in supporting measures to encourage the shift to electric and hasten the end of the internal combustion engine (ICE) vehicle. A considerable share of research funding has been directed towards battery development (primarily in the Horizon programme) and the electrification of utility and public transport vehicles. Many countries have provided substantial funding (through subsidy) to encourage citizens to buy electric vehicles, considering the higher cost of the vehicle when compared to ICE equivalents (Chapter 5). Some of these subsidies also feature as part of the EU Green Deal list of investments. The islands of Malta have been subsidising the purchase of new electric vehicles for the past few years, with the subsidy towards private electric vehicle ownership featured in the *Resilience and Recovery Plan for Malta* to the tune of €50.3 million (Government of Malta, 2023). The policy is linked to cities and urban areas, where the impact of transport generated pollution is higher.

Sales figures in Malta and beyond, however, remain challenging. The European Automobile Manufacturer's Association (ACEA, 2024) reports a 17.3% share of the EU new car market being battery-electric (an increase from 2023), although the volumes have dropped by 5.8% with the total market share falling to 13.1% from 14% in 2023. Plug-in hybrid car registrations in September 2024 have also declined by 22.3%. There is also the concern about the over-subsidisation of private vehicles, which continue to lock-in car dependence effects in cities and urban areas. Attard et al. (2015) warned about the low return on investment in terms of reduced externality costs when one considers air and noise pollution, congestion, accidents, and climate change adaptation overall costs. Indeed, the economic burden of congestion in many European cities is very high (Christidis & Ibáñez Rivas, 2012) and is a major concern which will not be resolved with the replacement of the fleet and the continued subsidisation for more private car purchases.

The move towards electrification also raises issues of energy sourcing (especially clean-energy systems), infrastructure for energy distribution (energy grids), and charging infrastructure in urban areas. The shift from fuel infrastructure to the electric grid can have impacts on the magnitude, timing, and location of electrical demand across a city (Elsayed et al., 2024). There are numerous studies that look at ways of ensuring a more effective electricity distribution, with analysis also split by mode and type of fleets. For example, Colovic et al. (2024) investigate the electricity required for freight transport and the potential of eHighways that could support electrification of the fleet, while Caroleo et al. (2024) develop a tool to guide strategies for an electric charging network for local public transport.

Given that the coupling of the transport and energy sectors is a relatively new concept, new models of operation and investments are being developed to combine, for example, public transport provision and electric bus fleet charging (Say et al., 2024). In many cases, the construction and ownership of expensive charging infrastructure in urban areas, operationalised within short concession contracts, deserves attention to ensure long-term sustainability of such infrastructures and their use. In some contexts where energy is imported, as in the case of islands, long-term plans are set in motion to ensure security of supply through interconnectors with the mainland. Malta's first interconnector with Sicily (completed in 2015) is being followed by a second interconnector, set to be operational in 2026, to ensure the demand of transport electrification can be supported (InterConnect Malta Ltd, 2023). Such projects also raise questions about long-term lock-in effects that are tied to fossil fuel use. The concerns over clean-energy systems are relevant where the potential of renewable energy production is limited. In the case of Malta, the limited land resources hinder considerably the growth of solar photovoltaic (PV) installations and the deep sea around its coastline (and its situation in the middle of the Mediterranean maritime routes) challenges the installation of offshore wind turbines. The building of an interconnector that delivers electricity from power that is generated from fossil fuel has been criticised for its long-term implications on decarbonisation efforts (Friends of the Earth Malta, 2024).

CONCLUSION

The future of urban transport is being continuously challenged by the needs of growing populations and infrastructural capacities which are being pushed to the limit. Attempts to resolve the problems associated with this growth of mobility through an increase in capacity (where this is still possible) has resulted in even greater congestion and pollution and in many cases lower quality of life for urban dwellers, especially those living close to roads and highways. These infrastructures are also costly to build and maintain with many cities struggling to keep up with the costs. Miller (2013) called out the need to identify new capabilities (rather than capacity) of the available transport infrastructure and increase efficiency and, potentially, capacity without the need to extend the infrastructure. He foresaw the opportunities arising from connected transport, use of geographic information systems and location-based services, and social media for citizens to share and collaborate, operating more collectively to solve mobility and accessibility problems. His work paved the way for many studies and eventually technologies that cities adopted to create multi-modal systems based on the simple principle of sharing resources. This is still relevant today with cities continuously struggling to develop their transport systems by enhancing 'legacy' systems such as public transport but also experimenting with new and more dynamic, shared transport modes.

The future of city transport systems will depend on how well we manage to integrate the pieces and develop strategies that complement each other, rather than work in silos and designing systems and infrastructures that serve only one mode (the car) and are not inclusive. Despite the many efforts of cities across Europe (e.g., the 15-minute city concept in Paris) and elsewhere (e.g., the TransMilenio BRT and cycling infrastructure in Bogotá), the 'system of automobility' as coined by John Urry (2013) is still very much alive and challenging the sustainability of cities across the world. If we are to really aspire for the goal of sustainable cities and communities (UN Sustainable Development Goal 11), then a more comprehensive approach is required which focuses on people and puts human well-being at the centre of design and decision-making.

This chapter identified and discussed aspects of urban transport which have the potential to impact our immediate and long-term future. Changes over the next few decades will be driven by population growth/decline, increased uncertainty due to climate change and a greater need for adaptation and resilience, and finally a drive towards improving overall living conditions in cities. The role of transport in all of this is central and both political and societal choices will impact the successful transition to decarbonisation and sustainable cities.

NOTES

1. Initiatives beyond Europe and the United Kingdom include the recent Mobilizing Justice partnership in Canada, aimed at addressing transport inequalities through research across different cities. https://mobilizingjustice.ca/.
2. See, for example, SKY ThinkTank Flying Zero. https://www.skythinktank.com/sky-thinktank-to-launch-maltas-first-drone-delivery-service-flying-zero.

REFERENCES

Aarhaug, J., Fearnley, N., Hartveit, K. J. L., & Johnsson, E. (2023). Price and competition in emerging shared e-scooter markets. *Research in Transportation Economics*, *98*, 101273.

Aarhaug, J., Fearnley, N., & Johnsson, E. (2023). E-scooters and public transport – Complement or competition? *Research in Transportation Economics*, *98*, 101273.

Aifandopoulou, G., & Xenou, E. (2019). *Sustainable urban logistics planning*. https://urban-mobility-observatory.transport.ec.europa.eu/document/download/9b248341-5a2e-4706-9dc2-5fa334fdcf58_en?filename=sustainable_urban_logistics_planning.pdf

Antoniou, C., Nocera, S., & Susilo, Y. (2023). The integration of passenger and freight transport: Trends, gaps and future research challenges. *Transportation Research Part A: Policy and Practice*, *173*, 103724.

ARUP, & NatCen. (2022). *National evaluation of e-scooter trials*. https://assets.publishing.service.gov.uk/government/uploads/system/uploads/attachment_data/file/1128454/national-evaluation-of-e-scooter-trials-findings-report.pdf

Attard, M. (2023). Climate change and urban logistics. In J. Monios, L. Budd, & S. Ison (Eds.), *The Routledge handbook of urban logistics* (Chapter 23, pp. 323–332). Routledge.

Attard, M., & Balbontin, C. (2024). Workshop 6 report: Micromobility movement in urban transport. *Research in Transportation Economics*, *103*, 101399.

Attard, M., Cañas, C., Scerri, K., Maas, S., & Giuliano, S. J. (2024). Nudging walking and cycling inn extreme car dependent communities. In M. Tira, M. Tiboni, M. Pezzagno, & G. Maternini (Eds.), *New challenges for sustainable urban mobility: Volume II* (pp. 123–131). Springer Nature.

Attard, M., Guzman, L., & Oviedo, D. (2023). Urban space distribution: The case for a more equitable mobility system. *Case Studies in Transport Policy*, *14*, 101096.

Attard, M., Thomopoulos, N., & Shiftan, Y. (2024). Conclusions: Working towards a sustainable ACT. In N. Thomopoulos, M. Attard, & Y. Shiftan (Eds.), *Sustainable automated and connected transport* (pp. 187–194). Emerald Publishing Limited.

Attard, M., Von Brockdorff, P., & Bezzina, F. (2015). *The external costs of passenger and commercial vehicle use in Malta*. University of Malta – European Commission Representation in Malta.

Beyazit, E., Soh, E., & Martens, K. (2024). Rescuing transport from inequities: How can automated and connected transport contribute to a more inclusive transport system? In N. Thomopoulos, M. Attard, & Y. Shiftan (Eds.), *Sustainable automated and connected transport* (pp. 139–160). Emerald Publishing Limited.

Birkenfeld, C., Victoriano-Habit, R., Alousi-Jones, M., Soliz, A., & El-Geneidy, A. (2023). Who is living a local lifestyle? Towards a better understanding of the 15-minute-city and 30-minute-city concepts from a behavioural perspective in Montréal, Canada. *Journal of Urban Mobility*, *3*, 100048.

Bjørgen, A., Seter, H., Kristiansen, T., & Pitera, K. (2019). The potential for coordinated logistics planning at the local level: A Norwegian in-depth study of public and private stakeholders. *Journal of Transport Geography*, *76*(4), 34–41.

Bruzzone, F. (2021). The integration of passenger and freight transport for first-last mile operations. *Transport Policy*, *100*, 31–48.

Büttner, B., Seisenberger, S., McCormick, B., Silva, C., Teixeira, J., Papa, E., & Cao, M. (2024). *Mapping of 15-minute city practices. Overview on strategies, policies and implementation in Europe and beyond*. Report from DUT's 15-minute city transitions pathway. https://dutpartnership.eu/news/new-publication-mapping of-15-minute-city-practices/

Büttner, B., Silva, C., Merlin, L., & Geurs, K. (2024). Just around the corner: Accessibility by proximity in the 15-minute city. *Journal of Urban Mobility*, *6*, 100095.

Cai, L., Yuen, K. F., Xie, D., Fang, M., & Wang, X. (2021). Consumer's usage of logistics technologies: Integration of habit into the unified theory of acceptance and use of technology. *Technology in Society*, *67*, 101789.

Caroleo, B., Lazzeroni, P., & Arnone, M. (2024). Towards full electrification of local public transport: A tool to guide strategies for implementing the electric charging network. *Journal of Urban Mobility*, *6*, 100088.

Cavallaro, F., & Nocera, S. (2022). Integration of passenger and freight transport: A concept-centric literature review. *Research in Transportation Business and Management*, *43*, 100718.

Chakravarty, A. K. (2014). *Supply chain transformations: Evolving with emerging business paradigms*. Springer.

Christidis, P., & Ibáñez Rivas, J. N. (2012). *Measuring road congestion*. European Commission Joint Research Centre. European Commission https://publications.jrc.ec.europa.eu/repository/handle/JRC69961

Colovic, A., Marinelli, M., & Ottomanelli, M. (2024). Towards the electrification of freight transport: A network design model of assessing the adoption of eHighways. *Transport Policy*, *150*, 106–120.

Combs, T. (2024). *Local actions to support walking and cycling during social distancing dataset*. http://pedbikeinfo.org/resources/resources_details.cfm?id=5209

Creutzig, F., Roy, J., Lamb, W. F., Azevedo, I. M. L., Bruine de Buin, W., Dalkmann, H., Edelenbosch, O. Y., Geels, F. W., Brubler, A., Hepburn, C., Hertwich, E. G., Khosla, R., Mattauch, L., Minx, J. C., Ramakrishnan, A., Rao, N. D., Steinberger, J. K., Tavoni, M., Ürge-Vorsatz, D., & Weber, E. U. (2018). Towards demand-side solutions from mitigating climate change. *Nature Climate Change*, *8*(4), 268–271.

Ding, X., & Jian, S. (2024). Revenue sharing and resource allocation for cooperative multimodal transport systems. *Transportation Research Part C: Emerging Technologies*, *164*, 104666.

D'Orso, G., Scerri, K., Attard, M., & Migliore, M. (2024, September 29–October 3). *Exploring factors influencing e-scooter use in islands*. Paper presented at 18th international conference on competition and ownership in land passenger transport (Thredbo 17), Cape Town, South Africa.

Elsayed, M., Zafar, K., Esa, Y., Soujad, T., & Mohamed, A. (2024). Impact of 100% vehicle electrification on the distribution grid in dense urban regions. *Energy Reports*, *11*, 5315–5322.

European Automobile Manufacturer's Association. (2024). *New car registration: -6.1% in September 2024; year-to-date battery electric market sales -5.8%*. European Automobile Manufacturer's Association. https://www.acea.auto/pc-registrations/new-car-registrations-6-1-in-september-2024-year-to-date-battery-electric-market-sales-5-8/

European Commission. (2013). *A call to action on urban logistics*. https://smartcities.at/wp-content/uploads/sites/3/A-call-to-action-on-urban-logistics.pdf

European Commission. (2019). *Malta: Summary of the Commission assessment of the draft national energy and climate plan 2021–2030*. https://energy.ec.europa.eu/document/download/b9bd0e66-595a-4e4c-af02-f32b112e3c96_en

European Commission. (2024). *Regulation (EU) 2024/1679 of the European Parliament and of the Council of 13 June 2024 on Union guidelines for the development of the trans-European transport network, amending Regulations (EU) 2021/1153 and (EU) No 913/2010 and repealing Regulation (EU) No 1315/2013*. https://publications.europa.eu/resource/cellar/cc3395a5-3516-11ef-b441-01aa75ed71a1.0006.03/DOC_1

European Environment Agency. (2015). *Noise briefing*. https://www.eea.europa.eu/soer/2015/europe/noise

European Environment Agency. (2024). *Transport and mobility*. https://www.eea.europa.eu/en/topics/in-depth/transport-and-mobility

European Union. (2017). *Sustainable urban mobility. European Parliament resolution of 2 December 2015 on sustainable urban mobility (2014/2242(INI)*. Official Journal of the European Union. https://eur-lex.europa.eu/legal-content/EN/TXT/?uri=celex:52015IP0423

Forbes Technology Council. (2023). *18 Smart-city technologies that will genuinely improve urban living*. https://www.forbes.com/councils/forbestechcouncil/2023/06/05/18-smart-city-technologies-that-will-genuinely-improve-urban-living/

Friends of the Earth Malta. (2024). *Fossil free Malta*. https://foemalta.org/fossil-free-malta/

Garau, C., Masala, F., & Pinna, F. (2016). Cagliari and smart urban mobility: Analysis and comparison. *Cities*, *56*, 35–46.

Geels, F. W. (2012). A socio-technical analysis of low-carbon transitions: Introducing the multi-level perspective into transport studies. *Journal of Transport Geography*, *24*, 471–482.

Gössling, S. (2016). Urban transport justice. *Journal of Transport Geography*, *54*, 1–9.

Gössling, S., Schröder, M., Späth, P., & Freytag, T. (2016). Urban space distribution and sustainable transport. *Transport Reviews*, *36*(5), 659–679.

Government of Malta. (2023). *Malta: Recovery and resilience plan*. Malta. https://fondi.eu/wp-content/uploads/2023/11/Malta-Recovery-and-Resilience-Plan-2023-7.11.2023.pdf

Hopkins, D., & Schwanen, T. (2021). Talking about automated vehicles: What do levels of automation do? *Technology in Society*, *64*, 101488.

InterConnect Malta Ltd. (2023). *Projects – 2nd Malta-Sicily cable link*. Government of Malta. https://icm.mt/projects/

International Energy Agency. (2021). *Net zero by 2050 hinges on a global push to increase energy efficiency*. International Energy Agency. https://www.iea.org/articles/net-zero-by-2050-hinges-on-a-global-push-to-increase-energy-efficiency

Lu, H., Zhang, K., Gkritza, K., Cai, H. (2021). Are shared electric scooters competing with buses? A case study in Indianapolis. *Transportation Research Part D: Transport and Environment*, *97*, 102877.

Lucas, K. (2012). Transport and social exclusion: Where are we now? *Transport Policy*, *20*, 105–113.

Lyons, G. (2018). Getting smart about urban mobility – Aligning the paradigms of smart and sustainable. *Transportation Research Part A: Policy and Practice*, *115*, 4–14.

Maas, S., Bugeja, M., & Attard, M. (2021). Sustainable tourism mobility in Malta: Encouraging a shift in tourist travel behaviour through an innovative smartphone app for trip planning. In L. Zamparini (Ed.), *Sustainable transport and tourism destinations* (Chapter 6). Emerald Publishing Limited.

McKinsey. (2023). *Commercial drone deliveries are demonstrating continued momentum in 2023*. https://www.mckinsey.com/industries/aerospace-and-defense/our-insights/future-air-mobility-blog/commercial-drone-deliveries-are-demonstrating-continued-momentum-in-2023

Miller, H. (2013). Beyond sharing: Cultivating cooperative transportation systems through geographic information science. *Journal of Transport Geography*, *31*, 296–308.

Mladenovic, L., Plevnik, A., & Rye, T. (2022). Implementing national support programmes for sustainable urban mobility plans in a multilevel governance context. *Case Studies on Transport Policy*, *10*, 1686–1694.

Moreno, C., Allam, Z., Chabaud, D., Gall, C., & Pratlong, F. (2021). Introducing the "15-Minute City": Sustainability, resilience and place identity in future post-pandemic cities. *Smart Cities*, *4*(1), 93–111.

Nello-Deakin, S. (2019). Is there such a thing as a 'fair' distribution of road space? *Journal of Urban Design*, *24*(5), 698–714.

Newman, P. W. G., & Kenworthy, J. R. (1996). The land use-transport connection: An overview. *Land Use Policy*, *13*(1), 1–22.

Nikitas, A. (2024). Can driverless transport be sustainable? A triple bottom line problematisation. In N. Thomopoulos, M. Attard, & Y. Shiftan (Eds.), *Sustainable automated and connected transport* (pp. 85–100). Emerald Publishing Limited.

Papa, E., & Lauwers, D. (2015). Smart mobility: Opportunity or threat to innovate places and cities? M. Schrenk, V. V. Popovich, P. Zeile, P. Elisei, & C. Beyer (Eds.), *Proceedings REAL CORP 2015* (pp. 543–550). Tagungsband.

Rupprecht Consult (Ed.). (2019). *Guidelines for developing and implementing a sustainable urban mobility plan* (2nd ed.). Rupprecht Consult.

Say, K., Csereklyel, Z., Brown, F. G., & Wang, C. (2024). The economics of public transport electrification: The charging dilemma. *Energy Economics*, *135*, 107648.

Social Exclusion Unit. (2003). *Making the connections: Final report on transport and social exclusion*. Office of the Deputy Prime Minister, London. https://www.ilo.org/media/312721/download

Stephan, F., Reinsperger, N., Grünthal, M., Paulicke, D., & Jahn, P. (2022). Human drone interaction in delivery of medical supplies: A scoping review of experimental studies. *PLoS One*, *17*, e0267664.

Tartaglia, M., & Petrozziello, E. (2024). Measuring the impact of institutional and territorial drivers for an efficient and smooth mobility-as-a-service (MaaS) implementation: A global analysis. In M. Tira, M. Tiboni, M. Pezzagno, & G. Maternini (Eds.), *New challenges for sustainable urban mobility: Volume II* (pp. 267–277). Springer Nature.

Thomopoulos, N., Attard, M., & Shiftan, Y. (2024). *Sustainable automated and connected transport*. Emerald Publishing.

Union Internationale des Transports Publics. (2017). *Urban public transport in the 21st century*. Statistics Brief. UITP. https://cms.uitp.org/wp/wp-content/uploads/2020/08/UITP_Statistic-Brief_national-PT-stats.pdf

United Nations. (2019). *World urbanization prospects: The 2018 revision, department of economic and social affairs, population division (ST/ESA/SER.A/420)*. New York.

United Nations. (2023). *The sustainable development goals report 2023: Special edition*. United Nations. https://unstats.un.org/sdgs/report/2023/The-Sustainable-Development-Goals-Report-2023.pdf

United Nations Human Settlements Programme. (2022). *World cities report 2022*. UN-Habitat. https://unhabitat.org/sites/default/files/2022/06/wcr_2022.pdf

Urry, J. (2013). *Societies beyond oil. Oil dregs and social futures*. Zed Books.

Wang Y., Wang Z., Hu X., Xue G., & Guan, X. (2022). Truck–drone hybrid routing problem with time-dependent road travel time. *Transportation Research Part C*, *144*, 103901.

Wygonik, E., & Goodchild, A. V. (2018). Urban form and last-mile goods movement: Factors affecting vehicle miles travelled and emissions. *Transportation Research Part D: Transport and Environment*, *61*, 217–229.

CHAPTER 15

TOWARDS LOW-CARBON MOBILITY IN RAPIDLY GROWING ECONOMIES: CHALLENGES AND LESSONS FROM LATIN AMERICA

Daniel Oviedo[a], Maria A. Wilches-Mogollon[a] and Natalia Villamizar Duarte[b]

[a]*University College London, UK*
[b]*Newcastle University, UK*

ABSTRACT

The transition to low-carbon transport in rapidly growing economies is a pivotal challenge in addressing global climate goals. Latin America, characterised by its urbanisation dynamics, informal transport systems, and stark socio-spatial inequalities, exemplifies the complexities and opportunities of this transition. This chapter reflects on such challenges and opportunities through various case studies in the region to explore the barriers and strategies for steering the region's transport systems towards sustainability. Key challenges include the dominance of car-oriented infrastructure, the marginalisation of informal and active transport modes, and the exclusion of vulnerable populations from planning processes. The environmental and social costs of these trends are significant, often reinforcing patterns of inequality while increasing greenhouse gas emissions. As cities like Bogotá, Santiago, and Belo Horizonte illustrate, these systemic barriers are not insurmountable but require integrated and inclusive approaches. By framing low-carbon mobility as a socio-environmental

Towards Transport Net Zero
Transport and Sustainability, Volume 20, 225–240

ISSN: 2044-9941/doi:10.1108/S2044-994120250000020015

imperative, this chapter highlights the need for policy implementation that aligns sustainability with equity, offering actionable insights for urban planners, policymakers, and local governments in Latin America and beyond.

Keywords: Low-carbon mobility; net zero; Latin America; social inclusion; urban resilience

INTRODUCTION

The transition to low-carbon mobility represents a critical juncture for rapidly growing economies facing the dual challenge of still-low-but-rising levels of private motorisation and the choice for purposefully developing a system anchored in public transport that supports fair and equitable access in proximity (Logan et al., 2022; Teoh et al., 2020). Latin America and the Caribbean (LAC) have experienced rapid urbanisation since the 1970s, with 82% of the region's population living in urban settlements (United Nations, 2018). Fast-paced changes in the region's demographics have led to changes in urban mobility systems, primarily driven by a strong emphasis on high-capacity public transport and improved non-motorised connections. This chapter reflects on the drivers of Latin America's everchanging urban mobility ecosystems and the innovations in transport policy and practice that have enabled the region to respond to some of its most salient challenges for access and connectivity, reflecting on their implications for the region's transition to low-carbon mobility. Low-carbon transitions in mobility are a significant contributing factor to the global push for net zero. In LAC and other contexts of the global south, this climate-responsive agenda also tackles persisting deficits in infrastructures and accessibility needs through clean technologies for urban transport and the fair, safe, and enjoyable use of active travel.

This chapter provides an overview of the main issues driving the unequal and challenging urban structures in which recent infrastructure and policy developments have sought to reverse decades of car-oriented planning, informal urbanisation, and urban expansion (Oviedo & Nieto-Combariza, 2021). We discuss how infrastructures, policies, and other innovations underpinned by big-picture policy and development targets, such as reducing transport-related carbon emissions, have shaped access to opportunities, defined urban experiences, and reproduced social inequities. In doing so, we recognise that, on the one hand, rapid demographic growth, sprawling urbanisation, and strained governance capacity have outpaced the ability of governments to respond adequately to increasing demands for more efficient and cleaner transport services. On the other hand, citizens' daily concerns, such as securing access to essential opportunities, do not necessarily align with climate-responsive choices and concepts that, compared with pressing needs, become foreign, as is the case with adopting net zero (Guzman et al., 2024). This mismatch between priorities and capacities has led to unequal distributions of the benefits and burdens of urban transport, disproportionately affecting already vulnerable populations, including women, children, older adults, and low-income groups (Oviedo & Guzman, 2021).

One consequence of the above mismatches and structural challenges for the transition is the configuration of urban transport systems across LAC, reflecting the persistent socio-spatial inequalities characterising its cities. Informal transport modes, such as minibuses and jitneys (aka *colectivos*) and motorcycle taxis, remain crucial for underserved populations but often operate with inefficient vehicles, contributing to emissions (Nieto-Combariza et al., 2024; see Chapter 8). At the same time, car-oriented policies, driven by rising aspirations for private vehicle ownership, have fostered a tendency from the wealthier segments of the population to avoid active travel as a mode of transport, reinforcing social and cultural expectations and stigma that relegate walking – and to some extent, cycling – as second- or third-class options for mobility (Delclòs-Alió et al., 2021). These dynamics not only exacerbate mobility-related inequalities beyond physical movement. They make it harder for society to widely adopt cleaner and healthier transport forms. Lengthy and costly commutes for those in peripheral areas and reduced access to social and economic opportunities, compounded by increasing exposure to carbon emissions, traffic risks, and social challenges such as crime and violence, further entrench existing cycles of disconnection and exclusion (Oviedo, 2021).

The challenges of reducing transport-related carbon emissions while improving accessibility, especially for the most vulnerable populations, in urban areas with a long history of fragmented planning systems and disconnected implementation processes are significant. Yet, governments in the LAC region, particularly during the last three decades, have innovated in infrastructure development and are using different material, fiscal, and behavioural mechanisms to incentivise more equitable and inclusive mobility in the region. The same challenges for urban access in LAC cities have offered unique opportunities for steering transitions to a more sustainable mobility ecosystem that is increasingly low carbon. Indeed, Latin America's cities provide valuable insights for integrating social equity into the low-carbon mobility transition. The region has pioneered innovative solutions, such as developing 2,044 km of bus rapid transit (BRT) lanes in systems across 64 cities (*Global BRTData*, 2024). Another example is the use of aerial cable cars, reaching 24.8% of the cable cars worldwide, as solutions for enabling access to high-capacity public transport systems in cities in Colombia, Mexico, and Venezuela, or as in the case of Bolivia, the consolidation of an aerial cable-car-based public transport network (Cardona-Urrea et al., 2024; Scholl et al., 2022). Finally, the consolidation and regional expansion of the *Ciclovía Recreativa* programmes, a programme for incentivising active mobility that has been localised throughout 374 cities in 17 LAC (Sarmiento et al., 2017), and the development of pop-up bike lanes during the COVID-19 pandemic, highlighting the potential for sustainable and equitable urban mobility. However, the path forward requires fundamentally rethinking transport systems to prioritise accessibility, equity, and sustainability.

Using accessibility – understood as 'the ease of reaching desired destinations given a number of available opportunities and intrinsic impedance to the resources used to travel from the origin to the destination' (Bocarejo & Oviedo, 2012, p. 146) – as a guiding principle offers a lens through which to evaluate and address disparities in transport systems. It shifts the focus from merely moving people to ensuring equitable access to opportunities, services, and spaces within urban environments. Accessibility metrics reveal the unequal distribution of transport benefits and highlight opportunities for policy interventions that prioritise the needs of vulnerable populations while advancing low-carbon mobility strategies (Scholl et al., 2022).

One salient point evident from the recent focus on accessibility is the need to consider short- and medium-distance travel options as an integral part of the transport ecosystem. Active mobility, such as walking and cycling, represents a cornerstone of low-carbon mobility strategies. Investments in infrastructure promoting active transport, such as segregated bike lanes and pedestrian zones, could lead to a reduction of carbon emissions, improvement of public health due to the achievement of recommended physical activity levels, and enhance urban inclusivity (Delclòs-Alió et al., 2021). Latin America has developed successful active transport initiatives and policies, especially in big cities such as Santiago and Bogotá, which underscore the potential of active mobility to contribute to both environmental and social goals (Rivas & Serebrisky, 2021).

Similarly, public transport systems such as those mentioned above and innovative local solutions for short- and mid-length trips, such as aerial cable cars, remain critical to achieving low-carbon mobility (Hidalgo et al., 2013; Morales-Betancourt et al., 2023). However, maximising their potential requires integrating them with land-use planning and ensuring affordability to prevent displacement and exclusion. Recent efforts towards electrification of bus fleets in the region in countries such as Chile, Brazil, and Colombia demonstrate political and financial commitment to a transition to cleaner and more efficient transport systems.

By engaging with communities, cities can co-create solutions that reflect the lived realities of diverse urban populations and adapt their strategies and timeframes to suit the multiple needs of these populations. Different examples across the region show how participatory approaches can address needs such as safety, care-related travel, and accessibility for low-income women (Guevara-Aladino et al., 2024). Furthermore, processes of formalisation and integration have, to an extent, considered the voices and participation of informal transport providers in the transition to cleaner transport (Bocarejo & Urrego, 2022). These initiatives highlight the importance of incorporating intersectional perspectives into transport planning.

The remainder of this chapter expands on the trends and strategies adopted by selected cities in the LAC region to reflect on their implications for transitioning to low-carbon mobility and the pathway to net zero. This chapter draws on various examples from across the region to highlight the role of technology, planning strategies, and policy innovations in addressing deep-seated challenges for mobility and access in more sustainable and inclusive manners. This chapter concludes with an analysis of the need for more inclusive and participatory planning that supports the localisation of these strategies.

THE CHALLENGES OF FRAGMENTED URBAN GROWTH

Latin America's urbanisation has been rapid and uneven, leading to a pattern of sprawling development that exacerbates socio-spatial divisions and inequalities in income and urban space distribution (Bösehans & Massola, 2018; Coatsworth, 2008; Ferrari et al., 2020). Urban expansion has outpaced infrastructure development, creating pockets of inaccessibility and exclusion (Useche et al., 2024). This has resulted in the concentration of opportunities in the city centres and in areas where people with higher purchasing power often inhabit, while low-income people have been pushed to the periphery in areas where resources, opportunities, and infrastructure are scarce (Obregón-Biosca, 2022; Oviedo, 2021; Ruiz-Padillo et al., 2022). This pattern is further intensified by the dominance of car-oriented infrastructure, which prioritises road expansion and private vehicle use over public and active transport investments (Delclòs-Alió et al., 2022). The result is a landscape where low-income populations often live far from employment centres and essential services, relying on lengthy and costly commutes.

These long commutes, paired with the increase of private cars and motorbikes, result in the increase of congestion levels, reaching points of 272 hours per year spent in congestion in Bogotá, implying financial, environmental, and health burdens for individuals and the cities, especially for the most vulnerable populations (Delclòs-Alió et al., 2022). Such disparities are not merely inconvenient – they reinforce cycles of poverty and social exclusion, where low-income residents have to rely heavily on walking and informal transport modes to navigate their environment, especially in cities such as Mexico City, Buenos Aires, Rio de Janeiro, and Sao Pablo where high public transport costs are unaffordable for 20% of the poorest households (Scholl et al., 2022). In this sense, active mobility is central in Latin America, especially in low-income populations, where walking to work or study is not a choice but a necessity (Marquet et al., 2017). In cities such as Bogotá, São Paulo, and Santiago de Chile, between 30% and 45% of trips made by low-income people are made on foot, in contrast to approximately 20% in high-income groups (Scholl et al., 2022). In addition to the higher frequency of walking trips by low-income groups, they also tend to travel longer distances (Luis De Moraes Ferrari et al., 2019). Furthermore, as a result of the segregation of the cities and the inequality of access to opportunities, these trips are often in peripheral regions with scarce transport infrastructure and resources, making walking the only transport mode, mainly due to a lack of choice and usually in precarious and unsafe conditions (Arellana, Alvarez, et al., 2021; Guzman et al., 2022; Oviedo Hernandez & Titheridge, 2016; Ureta, 2008).

In Latin America, unequal access to mobility infrastructure and services reflects broader social division. Women, for example, often face compounded barriers in accessing transport due to care responsibilities, safety concerns, and limited access to private vehicles. Similarly, low-income populations in peripheral areas endure higher costs of transport, longer travel times, and greater exposure to air pollution and traffic risks such as collisions, injuries, and death. Moreover, women in peripheral neighbourhoods, who often commute for two and three hours daily and are compelled to travel during early or late hours, face the sparsity of public transport services, heightening their vulnerability to gender-based violence (Libertun de Duren, 2020).

In this sense, policies that promote active modes and encourage the improvement of bicycle and walking infrastructure and aspects of the built environment also impact demographic and social changes, contributing to reducing inequalities that often characterise Latin American cities. Several studies have sought to identify the main barriers to urban infrastructure and the built environment that affect the adoption of active modes in LAC highlighting aspects such as road and citizen safety, the existence and quality of sidewalks and bike lanes, slope and connectivity of routes, weather conditions, and attractiveness (Arellana, Fuentes, et al., 2021; Bösehans & Massola, 2018; Gutiérrez et al., 2020; Larranaga et al., 2019; Lucchesi et al., 2018; Martínez-Buelvas et al., 2022; Pinhate et al., 2020; Rojas López et al., 2019; Ruiz-Padillo et al., 2022). Furthermore, several initiatives from regional international development organisations such as the Inter-American Development Bank (IDB) have strived to outline policy and practice guidelines for the development of more inclusive active mobility, albeit with a stronger focus on cycling in earlier studies (Scholl et al., 2022).

The lack of equitable access to urban mobility results in a phenomenon termed 'mobility poverty', which restricts individuals' participation in essential activities such as employment, education, and healthcare. This exclusion is compounded by traditional transport planning approaches that prioritise efficiency and demand over equity, overlooking the diverse needs of urban residents. These systemic issues underscore the need for frameworks that integrate social inclusion into mobility planning.

OPPORTUNITIES FOR A LOW-CARBON TRANSITION

Despite the challenges Latin American cities have endured in urbanisation and social inequalities, the region has a history of pioneering sustainable transport initiatives, including public transport innovations and active mobility policies. The quest for inclusive and sustainable cities has generated the need to rethink mobility and transportation policies in Latin America, with various studies focused on equity suggesting that transport planning must play a more direct role as a social policy (Oviedo & Guzman, 2020; Vecchio et al., 2020). Some of the most recognised examples are (1) the vast bicycle interventions both for promoting bicycle use for leisure and commuting in Bogotá and Mexico City; (2) the active travel promotion programmes developed in Bogotá and Santiago; (3) the development of BRT systems in Bogotá, Mexico City, and Santiago; (4) the implementation of aerial cable cars as part of the public transport system in Bogotá, La Paz, and Medellín; and (5) the participatory and inclusive city planning policies in Bogotá and Santiago.

Prioritising walking and cycling has been recognised in Latin American agendas as a mechanism to effectively address health, congestion, and urban inequality issues (IDB, 2022). The *ciclovías recreativas*, where streets are temporarily closed to cars for pedestrians and bicycle users to reclaim the streets and be physically active, have been widely implemented in Latin America (Mejia-Arbelaez et al., 2021). Several studies have shown that while the *ciclovías recreativas* are mainly leisure programmes, they contribute to promoting healthy lifestyles, reduce exposure to air pollution and street noise, and enhance the quality of life while democratising the space (Mejia-Arbelaez et al., 2021). Moreover, the implementation of *ciclovías recreativas* and the improvement of bicycle infrastructure has led urban inhabitants in Latin America to increase the use of bicycles as the main transport mode (Rosas-Satizábal & Rodriguez-Valencia, 2019; Wilches-Mogollon et al., 2024). One of the most recognised examples is Bogotá. The city has had the *Ciclovía Recreativa* since 1974, with an estimated 600,000–1,750,000 participants each Sunday, and 127.7 km in 2024 (Ciclovía Bogotana/IDRD – Instituto Distrital de Recreación y Deporte, 2020). The *ciclovía* in Bogotá and the several bicycle promotion policies and strategic plans that the city has experienced for over 30 years have led the city to have, nowadays, more than 880,000 bicycle trips per day over more than 630 km of segregated bicycle infrastructure (Rosas-Satizábal & Rodriguez-Valencia, 2019; Secretaría Distrital de Movilidad, 2019). The use of bicycles as a transport mode was especially important during the COVID-19 pandemic, forcing the different city governments to protect the population while reducing the economic impact of the reduction of economic activities. The implementation of 84 km of segregated pop-up bike lanes in Bogotá showcased how Latin American cities can innovate to generate safe spaces for rapidly growing transport demands while promoting low-carbon emission transport (Becker et al., 2022; Wilches-Mogollon et al., 2024).

Bogotá, Mexico City, and Quito have implemented bicycle promotion policies using leisure programmes to incentivise active travel adoption, further supported by capital investments in improving the connectivity and quality of their bicycle paths. Mexico City developed the 'Paseo Dominical Muévete en Bici' in 2007 to promote physical activity on Sundays (*Muévete en Bici*, 2012). Over the years, the programme has included a bicycle school, mechanical classes, inclusive trips, and physical activity classes throughout its 61 km of enclosed streets (*Muévete en Bici*, 2012). Similarly, the city accounts for 573.2 km of segregated bicycle infrastructure for everyday trips, with 37 km corresponding to pop-up bike lanes, with over 456,000 daily bicycle trips (Secretaría de Movilidad, 2024). Recent research in this context supports the role of such infrastructure and the development of shared bike systems as key elements in the transition of a potential 80% of car-based trips made within cyclable distances in the city (Oviedo & Sabogal-Cardona, 2022).

On the other hand, when analysing walking as a transport mode, the inequalities and lack of accessibility of LAC cities have led to walking being seen as a need, creating resistance to choosing walking as commute mode. In this sense, Santiago and Bogotá have developed active mobility to school programmes that aim to generate healthy lifestyles while promoting active mobility in a safe environment for the children and adolescents. In Santiago, the *Rutas bakanes* programme was developed using a participatory approach, where the children and adolescents were part of the co-creation of the programme, according to their needs (Sagaris & Lanfranco, 2019). Meanwhile Bogotá's active mobility programmes *Ciempiés Caminos Seguros* and *Al Colegio en Bici* were developed by the Mobility Secretariat 5 and 10 years ago, respectively, and have benefited children in more than 100 schools (Fabio Meneses Guerrero Cindy Yesenia Morales Espitia Asesora Natalia Pinilla Cortes, 2021; Hidalgo et al., 2016). Moreover, the pedestrian policy and the development plans, such as *Bogotá Camina Segura*, lead the way towards interventions that promote safe spaces for walking (Secretaría de Planeación, 2024).

BRT systems have played a fundamental role in improving urban mobility in dense and congested cities while reducing carbon emissions in Latin America. TransMilenio, one of the most recognised BRT systems in the world, was inaugurated in December 2000 in Bogotá. Since then, Buenos Aires, São Paulo, Lima, and Mexico City, among other Latin American cities, have developed BRT systems either to expand the capacity of their public transport systems or to integrate them. The implementation of BRT corridors in Buenos Aires led to a reduction of 53% of travel times and an increase of 105% in travel speed, decreasing carbon emissions as the population shifts from private vehicle users to public transport users (World Bank Group, 2024). Similarly, the renewal of the bus and BRT bus fleet in Bogotá has shown a 48% reduction in the in-bus eBC personal exposure, showing that updating the systems to the new energy trends can realise the benefits of BRTs in terms of public health and carbon emissions (Morales-Betancourt et al., 2022). In terms of accessibility, the BRT initiative in São Paulo doubled the number of jobs accessible within a 60-minute commute, while Lima's BRT has facilitated access to central city jobs (World Bank Group, 2024). These examples underscore the importance of BRT systems in reducing carbon emissions while improving accessibility.

Lastly, one of the innovations in motorised public transport was the implementation of aerial cable cars as part of the public transport systems in Latin America. More than 12 cities in Latin America use cable cars as feeders to their massive public transport systems, with Medellín implementing the first one in 2004 (World Bank Group, 2020). The cable cars in Medellín have delivered an increase in accessibility and safety perception (Cerdá et al., 2012; Dávila & Daste, 2005). The cable car in La Paz, at more than 30 km in length, has reduced the travel times by 22% while increasing the accessibility between the cities of La Paz and El Alto (Garsous et al., 2019). Bogotá is the latest Latin American city to implement an aerial cable car system. This implementation has been widely assessed, showing a significant reduction in the personal exposure and inhaled dose of air pollutants, increase in satisfaction with public transport, increase in access to parks, a perceived reduction in the neighbourhood's social stigma, increase in pride in community, and decrease in perceived insecurity (Baldovino-Chiquillo et al., 2023; Morales-Betancourt et al., 2023; Rubio et al., 2023). Aerial cable cars have shown their contribution to reducing carbon emissions in large cities with topographic barriers, while improving accessibility between the city centres and the peripheries, which could reduce the inequalities that urban settlements face in Latin America.

Cities' efforts to innovate in transport systems that can respond not only to a growing demand but also improve accessibility and reduce inequalities and carbon emissions demonstrate that targeted investments in sustainable infrastructure can have a broader set of positive outcomes than blanket approaches to transport improvements via technological or infrastructure developments. However, scaling such solutions requires a shift toward integrated and participatory planning approaches that prioritise social inclusion alongside environmental goals.

The key to the transition to low-carbon emissions is the redefinition of accessibility as a guiding principle, recognising that it encompasses more than physical proximity, thus ensuring equitable access to opportunities, services, and public spaces. This perspective aligns with the sustainable development goals' (SDGs') call for inclusive urban planning that addresses the needs of structurally disadvantaged groups, including women, children, and persons with disabilities. Integrating accessibility into transport planning can bridge the gap between environmental and social objectives, creating transport systems that are both low carbon and inclusive.

PARTICIPATORY APPROACHES AS MEANS OF FURTHER SUPPORTING THE TRANSITION IN EMERGING ECONOMIES

The pathway to net zero must consider the existing gaps in access and mobility across social and geographical urban structures. As highlighted by organisations such as the United Council for Local Governments (UCLG), participatory planning is essential to achieving a just transition to low-carbon mobility and account for urban residents' diverse needs and priorities, particularly those from marginalised communities (UCLG, 2023). These processes, underpinned by broader recognition of the interrelation between macro development objectives such as access and sustainability and structural drivers such as differences in social positions of individuals and their uneven power relations, enable the co-creation of solutions that reflect residents' lived experiences, fostering greater equity and political buy-in.

The transition to low-carbon mobility and access can thus be strengthened by mainstreaming care and inclusive perspectives into urban and territorial planning. As illustrated by recent experiences in different contexts, particularly in cities, implementing public care policies, such an approach requires the consideration of a lifecycle approach to addressing the safety and well-being of structurally disadvantaged groups. This involves enhancing infrastructure to meet their specific needs and aspirations. For example, Santa Fe, Argentina, faced challenges of discrimination and profound socio-economic inequalities that hindered low-income groups' access to essential services. The city's planning policies have embraced redesigning recreational and non-recreational public spaces through an inclusive feminist lens as a strategy to combat these structural inequities, enabling the appropriation of public spaces and the adoption of active travel for care and leisure mobilities at the local scale (UCLG, 2023).

Reimagining urban streets from children's perspective, for instance, highlights the potential to improve road safety and mobility for all users while embracing playing and active travel as critical pillars of local accessibility. Infrastructure upgrades should focus on meeting basic safety, accessibility, and mobility requirements, moderating vehicular speeds to prevent child fatalities, and extending the functionality of streets to include nearby spaces. Creating secure and engaging streetscapes contributes to an environment that supports children's cognitive development, educational success, and future societal contributions. In Quilmes, Argentina, the municipality has adopted a feminist approach to infrastructure planning aimed at fostering accessible public spaces and reducing gender disparities. A notable initiative involved designing murals with feminist themes, slogans like *Ni una menos*, and inclusive imagery to reclaim public spaces as hubs for community interaction, exchange, and participation (Secretaría de Mujeres y Diversidades, 2020).

Developing planning tools rooted in inclusion and mutual care principles can transform conventional practices by embracing intersectionality. Bogotá exemplifies this with its Care System, which seeks to build capacities, uphold women's rights, and involve men in caregiving responsibilities to address care labour inequities. Coordinated by the Secretariat for Women, this system integrates services from 13 sectors within the District Administration, enabling a co-responsible approach to caregiving. Feminist urbanism offers a framework for understanding how urban spaces shape women's and marginalised groups' lived experiences. This perspective emphasises designing urban environments that respond to everyday needs, fostering a sense of belonging and inclusion. Bogotá's 'Care Blocks' (*Manzanas del Cuidado*) (Guevara-Aladino et al., 2024; *Manzanas Del Cuidado – Tiempo y Servicios Para Las Mujeres*, 2024) are a notable example of redistributing caregiving responsibilities across governments, the private sector, communities, and households. These blocks provide educational support, rest, and well-being programmes at no cost, embodying the principles of 'caring cities and territories', where the urban environment supports those who care for others and their communities.

Moreover, participatory planning offers a pathway to integrate informal transport systems into formal frameworks. By engaging operators and users of informal modes in the planning process, cities can enhance efficiency, reduce emissions, and improve service quality without displacing those who rely on these systems. This approach addresses environmental goals and acknowledges informality's critical role in urban mobility, challenging the biases that have historically excluded these systems from sustainable development agendas. One of the most entrenched biases in transport and urban planning is the marginalisation of informality. As discussed earlier, informal transport systems are vital to urban mobility in many rapidly growing metropolitan regions across the Global South, with Latin America being no exception. These privately operated networks cater to the need for affordable and flexible transportation options, serving millions of daily commuters while employing thousands. Informal transport is critical in sustaining urban economies by connecting peripheral areas to city centres and addressing gaps left by formal systems. Despite its ubiquity, however, informal transport is often dismissed as a localised issue and excluded from broader urban and transport planning frameworks (Alcorn & Karner, 2023; Hidalgo & King, 2014; Rodriguez-Valencia et al., 2023).

This marginalisation is compounded by inconsistent and often discriminatory policies and regulations, which fail to acknowledge the sector's importance. Being a highly localised issue, local and regional governments have an opportunity to reverse this trend by actively engaging with operators of informal transport systems (see Chapter 8). Recognising the contributions of informal networks is essential to creating a more sustainable and inclusive urban mobility framework. Recent initiatives, such as those spearheaded by the United Nations Development Programme's (UNDP's) Accelerator Labs (UNDP, 2021), emphasise the strategic importance of informal mobility in urban settings. These efforts highlight the sector's critical role in urban economics and promote integration into global sustainable development agendas. The UNDP partnership seeks to deepen understanding of how various informal systems function and to explore the potential for hybrid models that balance formalisation with the flexibility of informal networks. This initiative involves testing innovative policy approaches, generating actionable data, and crafting specific policy recommendations to support the sector. By formalising these findings into institutional frameworks, the initiative aims to highlight opportunities within informal transport as a crucial element of sustainable urban development (UCLG, 2023).

CONCLUSION

LAC provide valuable lessons to reframe the transition to net zero as more than a technical challenge. Recognising that reducing carbon emissions and transport's environmental footprint, while maximising its role as an enabler of access to essential opportunities and the inclusive enjoyment of public spaces, is a socio-environmental imperative that demands integrated, inclusive, and participatory approaches. By addressing the structural inequalities embedded in the region's urban and transport systems, cities create mobility solutions that align sustainability with equity while advancing the transition to net zero. Latin America's experience offers valuable lessons for rapidly growing economies worldwide, demonstrating that the path to low-carbon mobility is not just about reducing emissions – it is about creating cities that are accessible, inclusive, and resilient for all their inhabitants.

The transition to low-carbon mobility in Latin America demands reimagining transport systems to align sustainability with equity, prioritising the needs of vulnerable populations. In this context, Latin America's cities hold the potential for equitable and sustainable urban futures regarding material, social, cultural, and behavioural solutions, fostering participatory planning and addressing systemic barriers to mobility. A step forward in scaling up these solutions is planning for their expansion considering the integration of both the productive (i.e., economic activities) and reproductive (i.e., care, family, social activities) spheres in urban and transport planning, with a focus on proximity that enable positive urban transformations supported by active travel. Such integration must give meaning to people's day-to-day lives across the schematic division between the city, public space, production, home, private space, and reproduction. Linking social services, such as education and urban health care, and implementing policies to prevent displacement and protect specific structurally discriminated groups advance social inclusion and contribute to a more equitable urban environment.

The LAC region has successfully developed material responses to its accessibility and mobility challenges, addressing some of its most salient social and spatial inequalities resulting from decades of centralised and car-oriented planning. The availability of a solid material backbone for sustainable urban mobility ecosystems such as the various high-capacity public transport, mid-sized transport infrastructures, and a growing network of facilities for non-motorised mobility must be strengthened by strategic partnerships for more inclusive and participatory planning. For instance, by partnering with informal transport operators and communities, it is possible to co-produce systems prioritising vulnerable populations' mobility needs. Such collaborations can lay the foundation for sustainable and inclusive urban transport systems that enhance access to opportunities for all residents while challenging traditional planning paradigms that have often overlooked or stigmatised informality.

The lessons from the region underscore that achieving low-carbon mobility is inseparable from the broader goals of social inclusion and urban resilience in the global south. The main takeaway from our chapter is the need for more inclusive and participatory planning. This approach will support the broader adoption of often-isolated local strategies (i.e., innovations that remain confined to a given city or neighbourhood) through context-specific adaptation and the empowerment of local and regional governments to work more closely with communities and disadvantaged actors. Furthermore, adopting a more participatory approach to the transport system's supply and demand sides will foster social innovation and develop bottom-up, community-led – or at least community-oriented – low-carbon mobility solutions.

REFERENCES

Alcorn, L. G., & Karner, A. (2023). From minibuses to "boda bodas," informal transport could be an untapped climate change solution. *Transportation*, *48*(3), 1361–1377.

Arellana, J., Alvarez, V., Oviedo, D., & Guzman, L. A. (2021). Walk this way: Pedestrian accessibility and equity in Barranquilla and Soledad, Colombia. *Research in Transportation Economics*, *86*, 101024.

Arellana, J., Fuentes, L., Cantillo, J., & Alvarez, V. (2021). Multivariate analysis of user perceptions about the serviceability of urban roads: Case of Barranquilla. *International Journal of Pavement Engineering*, *22*(1), 54–63.

Baldovino-Chiquillo, L., Sarmiento, O. L., Phd, D., Wilches-Mogollon, M. A., Aguilar, A. F., Brainlat, Javeriana, U., Florez-Pregonero, A., Sarmiento, O. L., Baldovino-Chiquillo, L., O'donovan, G., Wilches-Mogollon, M. A., Aguilar, A. F., Florez-Pregonero, A., Martínez, P. A., Arellana, J., Guzmán, L. A., Yamada, G., Rodriguez, D. A., & Diez-Roux, A. V. (2023). Effects of an urban cable car intervention on physical activity: The TrUST natural experiment in Bogotá, Colombia. *The Lancet Global Health*, *11*(8), e1290–e1300.

Becker, S., von Schneidemesser, D., Caseiro, A., Götting, K., Schmitz, S., & von Schneidemesser, E. (2022). Pop-up cycling infrastructure as a niche innovation for sustainable transportation in European cities: An inter- and transdisciplinary case study of Berlin. *Sustainable Cities and Society*, *87*, 104168.

Bocarejo, J. P., & Urrego, L. F. (2022). The impacts of formalization and integration of public transport in social equity: The case of Bogota. *Research in Transportation Business & Management*, *42*, 100560.

Bocarejo, S. J. P., & Oviedo H., D. R. (2012). Transport accessibility and social inequities: A tool for identification of mobility needs and evaluation of transport investments. *Journal of Transport Geography*, *24*, 142–154.

Bösehans, G., & Massola, G. M. (2018). Commuter cyclists' risk perceptions and behaviour in the city of São Paulo. *Transportation Research Part F: Traffic Psychology and Behaviour*, *58*, 414–430.

Cardona-Urrea, S., Soza-Parra, J., & Ettema, D. (2024). Aerial cable cars as a transit mode: A review of technological advances, service area characteristics, and societal impacts in Latin America and the Caribbean. *Transport Reviews*, *44*(3), 684–708.

Cerdá, M., Morenoff, J. D., Hansen, B. B., Tessari Hicks, K. J., Duque, L. F., Restrepo, A., & Diez-Roux, A. V. (2012). Reducing violence by transforming neighborhoods: A natural experiment in Medellín, Colombia. *American Journal of Epidemiology*, *175*(10), 1045–1053.

Ciclovía Bogotana/IDRD – Instituto Distrital de Recreación y Deporte. (2020). Retrieved December 9, 2024, from https://www.idrd.gov.co/recreacion/ciclovia-bogotana

Coatsworth, J. H. (2008). Inequality, institutions and economic growth in Latin America. *Journal of Latin American Studies*, *40*(3), 545–569.

Dávila, B. J. D., & Daste, D. (2005). *Medellín's aerial cable-cars: Social inclusion and reduced emissions*. University College London.

Delclòs-Alió, X., Rodríguez, D. A., Medina, C., Miranda, J. J., Avila-Palencia, I., Targa, F., Moran, M. R., Sarmiento, O. L., & Quistberg, D. A. (2021). Walking for transportation in large Latin American cities: Walking-only trips and total walking events and their sociodemographic correlates. *Transport Reviews*, *42*(3), 296.

Delclòs-Alió, X., Rodríguez, D. A., Olmedo, N. L., Ferrer, C. P., Moore, K., Stern, D., de Menezes, M. C., de Oliveira Cardoso, L., Wang, X., Guimaraes, J. M. N., Miranda, J. J., & Sarmiento, O. L. (2022). Is city-level travel time by car associated with individual obesity or diabetes in Latin American cities? Evidence from 178 cities in the SALURBAL project. *Cities*, *131*, 103899.

Fabio Meneses Guerrero Cindy Yesenia Morales Espitia Asesora Natalia Pinilla Cortes, D. (2021). *Ciempiés Caminos Seguros: Una apuesta para el fomento de estilos saludables*.

Ferrari, G. L. de M., Kovalskys, I., Fisberg, M., Gómez Salas, G., Rigotti, A., Cortés Sanabria, L. Y., Yépez García, M. C., Pareja Torres, R. G., Herrera-Cuenca, M., Zimberg, I. Z., Guajardo, V., Pratt, M., Gonçalves, P. B., Rosales-Salas, J., Cristi-Montero, C., Rodríguez-Rodríguez, F., Waddell, H., Petermann-Rocha, F., Celis-Morales, C. A., … Solé, D. (2020). Socio-demographic patterns of public, private and active travel in Latin America: Cross-sectional findings from the ELANS study. *Journal of Transport & Health*, *16*, 100788.

Garsous, G., Suárez-Alemán, A., & Serebrisky, T. (2019). Cable cars in urban transport: Travel time savings from La Paz-El Alto (Bolivia). *Transport Policy*, *75*, 171–182.

Global BRTData. (2024). Retrieved December 9, 2024, from https://brtdata.org/

Guevara-Aladino, P., Sarmiento, O. L., Rubio, M. A., Gómez-García, L. M., Doueiri, Z. N., Martínez, D., King, A. C., Hurtado-Tarazona, A., Banchoff, A., Guzman, L. A., Álvarez-Rivadulla, M. J., & Palencia, L. (2024). Urban care for unpaid caregivers: Community voices in the care block program, in Bogotá, Colombia. *Journal of Urban Health*, *2024*, 1–15.

Gutiérrez, M., Hurtubia, R., & Ortúzar, J. d. D. (2020). The role of habit and the built environment in the willingness to commute by bicycle. *Travel Behaviour and Society*, *20*(October 2019), 62–73.

Guzman, L. A., Arellana, J., & Castro, W. F. (2022). Desirable streets for pedestrians: Using a street-level index to assess walkability. *Transportation Research Part D: Transport and Environment*, *111*, 103462.

Guzman, L. A., Oviedo, D., & Cantillo-Garcia, V. A. (2024). Is proximity enough? A critical analysis of a 15-minute city considering individual perceptions. *Cities*, *148*, 104882.

Hidalgo, D., & King, R. (2014). Public transport integration in Bogotá and Cali, Colombia – Facing transition from semi-deregulated services to full regulation citywide. *Research in Transportation Economics*, *48*, 166–175.

Hidalgo, D., Miranda, L., Lleras, N., & Ríos, J. (2016). Al Colegio en Bici: Bike-to-school program in Bogotá, Colombia. *Transportation Research Record*, *2581*, 66–70.

Hidalgo, D., Pereira, L., Estupiñán, N., & Jiménez, P. L. (2013). TransMilenio BRT system in Bogota, high performance and positive impact – Main results of an ex-post evaluation. *Research in Transportation Economics*, *39*(1), 133–138.

IDB. (2022). *Cities on pedals: The IDB Group's commitment to cycling infrastructure*. Inter-American Development Bank. Retrieved January 30, 2025 from https://blogs.iadb.org/ciudades-sostenibles/en/bicycle-day-the-idb-group-commitment-to-cycling-infrastructure-bike-lane/

Larranaga, A. M., Arellana, J., Rizzi, L. I., Strambi, O., & Cybis, H. B. B. (2019). Using best–worst scaling to identify barriers to walkability: A study of Porto Alegre, Brazil. *Transportation*, *46*(6), 2347–2379.

Libertun de Duren, N. R. (2020). Effects of neighborhood upgrading programs on domestic violence in Bolivia. *World Development Perspectives*, *19*, 100231. https://doi.org/10.1016/j.wdp.2020.100231

Logan, K. G., Hastings, A., & Nelson, J. D. (2022). *Transportation in a net zero world: Transitioning towards low carbon public transport*. Springer.

Lucchesi, S. T., Uriarte, A. M. L., & Cybis, H. B. B. (2018). Aplicação da teoria de preços hedônicos para avaliação da influência da caminhabilidade no preço de venda de imóveis residenciais. *Transportes*, *26*(3), 120–133.

Luis De Moraes Ferrari, G., Kovalskys, I., Fisberg, M., Gómez, G., Rigotti, A., Sanabria, L. Y. C., García, M. C. Y., Torres, R. G. P., Herrera-Cuenca, M., Zimberg, I. Z., Guajardo, V., Pratt, M., King, A. C., & Solé, D. (2019). Original research Socio-demographic patterning of self-reported physical activity and sitting time in Latin American countries: Findings from ELANS. *BMC Public Health*, *19*(1), 1–12.

Manzanas del Cuidado – Tiempo y servicios para las mujeres. (2024). Retrieved December 9, 2024, from https://manzanasdelcuidado.gov.co/

Marquet, O., Hipp, J. A., & Miralles-Guasch, C. (2017). Neighborhood walkability and active ageing: A difference in differences assessment of active transportation over ten years. *Journal of Transport & Health*, *7*, 190–201.

Martínez-Buelvas, L., Rakotonirainy, A., Grant-Smith, D., & Oviedo-Trespalacios, O. (2022). A transport justice approach to integrating vulnerable road users with automated vehicles. *Transportation Research Part D: Transport and Environment*, *113*, 103499.

Mejia-Arbelaez, C., Sarmiento, O. L., Vega, R. M., Castillo, M. F., Truffello, R., Martínez, L., Medina, C., Guaje, O., Ortiz, J. D. P., Useche, A. F., Rojas-Rueda, D., & Delclòs-Alió, X. (2021). Social inclusion and physical activity in Ciclovía Recreativa programs in Latin America. *International Journal of Environmental Research and Public Health*, *18*(2), 655.

Morales-Betancourt, R., Galvis, B., Mendez-Molano, D., Rincón-Riveros, J. M., Contreras, Y., Montejo, T. A., Rojas-Neisa, D. R., & Casas, O. (2022). Toward cleaner transport alternatives: Reduction in exposure to air pollutants in a mass public transport. *Environmental Science & Technology*, *56*(11), 7096–7106.

Morales-Betancourt, R., Wilches-Mogollon, M. A., Sarmiento, O. L., Mendez-Molano, D., Angulo, D., Filigrana, P., Arellana, J., Guzman, L. A., Garzon, G., Gouveia, N., Levy, P., & Diez-Roux, A. V. (2023). Commuter's personal exposure to air pollutants after the implementation of a cable car for public transport: Results of the natural experiment TrUST. *Science of The Total Environment*, *865*, 160880.

Muévete en Bici. (2012). Retrieved December 9, 2024, from http://data.sedema.cdmx.gob.mx/mueveteenbici/index.php?option=com_content&view=article&id=44&it

Nieto-Combariza, M., San Gil, A., Quesada, A., Agudelo, D., Arellana, J., & Oviedo, D. (2024). Motorized three-wheelers and their potential for just mobility in Caribbean urban areas. *Data & Policy*, *6*, e11.

Obregón-Biosca, S. A. (2022). Choice of transport in urban and periurban zones in metropolitan area. *Journal of Transport Geography*, *100*, 103331.

Oviedo, D. (2021). Making the links between accessibility, social and spatial inequality, and social exclusion: A framework for cities in Latin America. *Advances in Transport Policy and Planning*, *8*, 135–172.

Oviedo, D., & Guzman, L. (2020). Transportation planning and development in Bogotá: Balancing the urgent and the strategic. In D. Rukmana (Ed.), *The Routledge handbook of planning megacities in the global south* (pp. 133–146). Routledge.

Oviedo, D., & Nieto-Combariza, M. (2021). Transport planning in the global south. In *International encyclopedia of transportation* (Vol. 7, pp. 118–123). Elsevier.

Oviedo, D., & Sabogal-Cardona, O. (2022). Arguments for cycling as a mechanism for sustainable modal shifts in Bogotá. *Journal of Transport Geography*, *99*, 103291.

Oviedo Hernandez, D., & Titheridge, H. (2016). Mobilities of the periphery: Informality, access and social exclusion in the urban fringe in Colombia. *Journal of Transport Geography*, *55*, 152–164.

Pinhate, T. B., Parsons, M., Fisher, K., Crease, R. P., & Baars, R. (2020). A crack in the automobility regime? Exploring the transition of São Paulo to sustainable urban mobility. *Cities*, *107*, 102914.

Rivas, M. E., & Serebrisky, T. (2021). *The role of active transport modes in enhancing the mobility of low-income people in Latin America and the Caribbean*. Inter-American Development Bank. https://doi.org/10.18235/0003216.

Rodriguez-Valencia, A., Rosas-Satizábal, D., & Hidalgo, D. (2023). Big effort, little gain for users: Lessons from the public transport system reform in Bogotá. *Public Transport*, *15*(2), 411–433.

Rojas López, M. C., Recalde Melo, D. F., & Wong, Y. D. (2019). Transport barriers and its health implications in Asunción. *Journal of Transport & Health*, *14*, 100579.

Rosas-Satizábal, D., & Rodriguez-Valencia, A. (2019). Factors and policies explaining the emergence of the bicycle commuter in Bogotá. *Case Studies on Transport Policy*, *7*(1), 138–149.

Rubio, M. A., Sarmiento, O. L., Guevara, T., O'Donovan, G., Muñoz, A., Guzman, L. A., Arellana, J., Molano, L., Huertas, B., Indvik, K., Segura, E., King, A. C., & Diez-Roux, A. V. (2023). The impacts of an urban cable car system on liveability: a mixed methods study in Bogotá, Colombia. In N. Biloria, G. Sebag, & H. Robertson (Eds.), *The Empathic City: An Urban Health and Wellbeing Perspective* (pp. 311–345). Springer Nature Switzerland, 2023.

Ruiz-Padillo, A., Oestreich, L., Torres, T. B., Rhoden, P. S., Larranaga, A. M., & Cybis, H. B. (2022). Weighted assessment of barriers to walking in small cities: A Brazilian case. *Transportation Research Part D: Transport and Environment*, *109*, 103392.

Sagaris, L., & Lanfranco, D. (2019). Beyond 'safe': Chilean 'Kool' routes to school address social determinants of health. *Journal of Transport & Health*, *15*, 100665.

Sarmiento, O. L., Díaz del Castillo, A., Triana, C. A., Acevedo, M. J., Gonzalez, S. A., & Pratt, M. (2017). Reclaiming the streets for people: Insights from Ciclovías Recreativas in Latin America. *Preventive Medicine*, *103*, S34–S40.

Scholl, L., Fook, A., Rebolledo, J. D. B., Rivas, M. E., Montes, L., Montoya, V., Pedraza, L., Noboa, N., Sandoval, D., Lee, S., Porcel, M. R., Bocarejo, J. P., Tovar, E. V., Urrego, L. F., Moreno, J. P., Bertucci, J. P., Oviedo, D., Sabogal-Cardona, O., Serebrisky, T., & Mojica, C. (2022). *Transport for inclusive development: Defining a path for Latin America and the Caribbean*. Inter-American Development Bank. https://doi.org/10.18235/0004335

Secretaría de Movilidad. (2024). *Conteo Ciclistas 2024*.

Secretaría de Mujeres y Diversidades. (2020). *Secretaría de Mujeres y Diversidades: Plan municipal para la Igualdad de Géneros*.

Secretaría de Planeación. (2024). *Bogotá Camina Segura: Plan Distrital de Desarrollo 2024-2028*.

Secretaría Distrital de Movilidad. (2019). *Encuesta de Movilidad 2019*.

Teoh, R., Anciaes, P., & Jones, P. (2020). Urban mobility transitions through GDP growth: Policy choices facing cities in developing countries. *Journal of Transport Geography*, *88*, 102832.

United Council for Local Governments. (2023). *Integrated and participatory urban planning: How local and regional governments enable equality through feminism, accessibility and proximity*. UCLG.

United Nations. (2018). *World urbanization prospects – Population division*. Retrieved December 8, 2024, from https://population.un.org/wup/

United Nations Development Programme. (2021). *Informal transportation: A new research agenda for the UNDP Accelerator Labs, in partnership with NewCities' global partnership for informal transportation*. United Nations Development Programme. Retrieved December 9, 2024, from https://www.undp.org/acceleratorlabs/press-releases/informal-transportation-new-research-agenda-undp-accelerator-labs-partnership-newcities-global-partnership-informal

Ureta, S. (2008). To move or not to move? Social exclusion, accessibility and daily mobility among the low-income population in Santiago, Chile. *Mobilities*, *3*(2), 269–289.

Useche, A. F., Sarmiento, O. L., Álvarez-Rivadulla, M. J., Medina, P., Higuera-Mendieta, D., & Montes, F. (2024). Spatial segregation patterns and association with built environment features in Colombian cities. *Cities*, *152*, 105217.

Vecchio, G., Tiznado-Aitken, I., & Hurtubia, R. (2020). Transport and equity in Latin America: A critical review of socially oriented accessibility assessments. *Transport Reviews*, *40*(3), 354–381.

Wilches-Mogollon, M. A., Sarmiento, O. L., Medaglia, A. L., Montes, F., Guzman, L. A., Sánchez-Silva, M., Hidalgo, D., Parra, K., Useche, A. F., Meisel, J. D., Ochoa-Montero, H., & Rodríguez Castañeda, N. (2024). Impact assessment of an active transport intervention via systems analytics. *Transportation Research Part D: Transport and Environment*, *128*, 104112.

World Bank Group. (2020). *Urban aerial cable cars as mass transit systems: Case studies, technical specifications, and business models*. World Bank Group.

World Bank Group. (2024). *Replicable mass transit systems reduce emissions and connect people to opportunities in Latin American cities*. https://www.worldbank.org/en/results/2024/04/23/replicable-mass-transit-systems-reduce-emissions-and-connect-people-to-opportunities-in-latin-american-cities

CHAPTER 16

GENDER AND THE TRANSITION TO NET ZERO TRANSPORT

Lucy Budd[a,b]

[a]*De Montfort University, UK*
[b]*Institute of Transport and Logistics Studies, The University of Sydney, Australia*

ABSTRACT

The United Nations has identified climate change and inequality – including gender inequality – as being two of the biggest challenges of the contemporary global era. The global transport sector, which currently accounts for around a quarter of all anthropogenic greenhouse gas emissions, remains a long way from reaching its net zero target and largely fails to deliver gender inclusive services which recognise and meet the different mobility needs, priorities, and services both between and within genders. The aim of this chapter is to review the current research landscape into the emerging and complex intersections between gender and the moves towards net zero transport. To do so, it will address issues of transport decarbonisation as it pertains to gender, before attending to pressing issues relating to the potential effects of proposed decarbonisation interventions on women workers in the transport sector and women users of transport systems. Current challenges and opportunities for change are discussed and an agenda for future gender-just net zero transport transitions proposed.

Keywords: Gender; gender equality; women; transport; net zero; gender-just transitions

Towards Transport Net Zero
Transport and Sustainability, Volume 20, 241–254

ISSN: 2044-9941/doi:10.1108/S2044-99412025000020016

INTRODUCTION

Climate change and inequality, including gender inequality, have been identified by the United Nations as two of the biggest challenges of the contemporary global era. Although much has been made both of the need to achieve net zero carbon emissions to mitigate the worst extremes of climate chaos and the policies and technologies which may or may not facilitate a transition to a dramatically carbon-curtailed (if not wholly carbon emission-free) future, it is only relatively recently that concerns about the gender and gendered implications of the move towards net zero transport have begun to be articulated. This is an important development. The global transport sector, which currently accounts for around a quarter of all anthropogenic greenhouse gas emissions, remains a long way from reaching its net zero target (International Transport Forum (ITF), 2021). In addition, transport has historically failed to deliver adequate gender inclusive services which recognise and meet the different mobility needs, priorities, and services both between and within genders and which also take account of changes in mobility patterns over the life course. Given that women's mobility, on average, has a lower carbon intensity than that of men (Polk, 2003) and women generally exhibit more environmentally aware behaviours and attitudes (Desrochers et al., 2019; Rosqvist, 2019), women can be powerful agents for accelerating transport decarbonisation (Kronsell et al., 2016). However, a gender-just change to decarbonise transport, which recognises and considers the implications of any intervention, can only occur if women are in positions to inform and influence investment and decision-making processes pertaining to the provision of future mobility-related issues, including the design and delivery of future infrastructure, vehicles, and services. As Lyons (2004, p. 494) noted 'The majority of the population is female and yet … the transport systems have been shaped around male, middle-class, middle-aged professionals …'. Over 20 years later, it could be argued that little has changed.

The ITF (2022, p. 8) is among a number of major organisations which have recognised that 'Gender equality, transport and climate change have for far too long been treated as separate problems that … [were] … addressed with different approaches'. Growing awareness of the interconnected nature of gender, transport, and climate change is now generating new conversations and ideas about how future transport policy, planning, and provision in a decarbonised world can help address deep-rooted structural and cultural barriers to mobility and deliver more equitable and sustainable transport outcomes for all in society. This includes women, people with other gender identities, people of all races, children, the elderly, and those with disabilities who may have found themselves previously marginalised by transport systems that were not adequately meeting their needs. By bringing different perspectives, social norms, knowledge, and experiences to the development of transport policy and provision, the diverse voices of women – as transport users, transport workers, and transport decision-makers – have the potential to deliver more gender equitable and environmentally sustainable transport. This, it is argued, will not only benefit women but also contribute to global economic development by enhancing women's access to education and participation in the workforce. In the words of a World Bank Report:

> Lack of adequate transport often translates into girls missing school, women limiting themselves to job opportunities closer to home, unemployment, and the inability to access health or childcare services. If women had full equality with men in choosing and using transport, they would have more control over their life choices. Enhancing women's mobility could also have a transformative impact on the global economy through increased labor force participation. (World Bank, 2024, unpaginated)

Within the context of current transport decarbonisation debates, there is also a need to ensure that women are not harmed or disadvantaged by the transition to net zero but actively benefit from the process and the opportunities it presents. The starting point for these new conversations around climate change/net zero, gender, and transport often coalesces around three inconvenient but incontrovertible truths:

1. *Transport remains a major global contributor of carbon dioxide and other potent greenhouse gas emissions*. It is also one of the most challenging sectors to decarbonise on account of its reliance on legacy infrastructure and vehicle fleets which predominately still rely on fossil fuels. According to the International Energy Agency (IEA, 2024), the global transport sector continues to rely on oil products to deliver 91% of its final energy needs, and this, combined with current demand trajectories, means the sectors are considered 'not on track' to meet its 2050 net zero target.
2. *Transport remains a male-dominated sector*. It is estimated that only 17% of transport workers worldwide are women. Although this figure is higher in some nations (for example, 26% in the United Kingdom) (Francis-Devine & Hutton, 2024), women remain under-represented in senior leadership (McLaughlin & Fearon, 2022) and safety-critical roles (McCarthy et al., 2015) and over-represented in poorer-paid, less secure and less societally respected roles in catering, cleaning, secretarial and customer services (Francis-Devine & Hutton, 2024). Consequently, the opportunities for women working in transport to change policy and practice and contribute to decarbonising the sector are, at present, limited.
3. Issues of transport affordability, accessibility, availability, cultural acceptability, and personal safety and security, when combined with their differing roles in society, mean *the transport choices (and thus mobility patterns) of men and women differ*, often quite significantly. A large corpus of academic literature has demonstrated that the mobility of women and men worldwide is affected by their economic and cultural position in society and that an individual's access to, decision to use, and experiences of using, different forms of public and private transport are impacted by multiple intersecting factors relating to their gender, age, wealth, dis/abilities, caring responsibilities, residential location, income, fear of crime, and concerns for personal safety (see Borgato et al., 2021; Carver & Veitch, 2020; Ceccato & Loukaitou-Sideris, 2022; d'Agostino et al., 2024). A failure to recognise these differences in mobility patterns and

practices could mean that net zero interventions which are solely designed to decarbonise the transport sector and which do not consider gender may have unintended and undesirable consequences for women and female mobility and/or are not as effective as they might be in terms of decarbonisation. The ITF's (2022) call to recognise the gender(ed) dimension that any changes in transport policy and provision have for women using, working in, and relying on the transport sector thus appears compelling.

Taking these three issues as its starting point, the aim of this chapter is to review the current research landscape into some of the emerging and complex intersections between gender and the moves towards net zero transport. To do so, it will address these points in turn, beginning with issues of transport decarbonisation as it pertains to gender, before attending to pressing issues relating to the potential effects of proposed decarbonisation interventions on women working in the transport sector (including the wider supply chain) and women as users of transport systems. In each case, current challenges and opportunities for change will be discussed before this chapter concludes by proposing an agenda for gender-just transport transitions and signposting to areas of future research need.

DECARBONISING TRANSPORT – A GENDER PERSPECTIVE

In a 2023 paper in the journal *Energy Research and Social Science*, geographer Debbie Hopkins and colleagues discussed the manifestation of what they termed 'Aeromasculinities' and the 'fallacy of sustainable aviation'. Taking what is arguably the most environmentally damaging and visible mode of rapid international mobility and long-distance travel as their starting point (see Chapter 10), they expose, through a critical analysis of the environmental statements of major global airlines, a 'gendered system of thinking, being and doing which forecloses radical action' and which prevents the 'change required for a climate-safe and just energy future' (Hopkins et al., 2023, p. 2). Noting that 95% of airline CEOs are male and that dominant narratives of 'sustainable aviation' continue to be situated within a global system of aeromobilities which was both created and sustained by fossil fuel dependence, they suggest that 'resisting aeromasculinities is not about 'adding women' to aviation boards (as reported in corporate social responsibility documents) or 'thinking about women' in constructing aeromobile futures but about the 'simultaneous challenging of heteronormative, white masculinities and responses to ecological collapse and the climate crisis' (Hopkins et al., 2023). Their provocation that environmentalism within transport (and society at large) is inherently gendered and that airlines' corporate environmental concerns and commitments can similarly be analysed through a gendered lens is both challenging and pertinent to the present discussion.

Scholars, activists, and commentators have long recognised the gendered (and also often racialised) cultures of aviation which reproduce and reinforce dominant technical-social logics, regimes, and bodies in flight (see Hopkins et al., 2023, and also Adey et al., 2024). From initial refusals to appoint women as cabin crew in the 1920s, through to overt sexualisation of female flight attendants in the 1960s and 1970s and explicit discrimination against women pilots in the 1980s and 1990s (see Smith et al., 2021), to the low numbers of women holding senior (executive or board level) management positions at airports in the 2020s (Amaugo, 2022), the history of commercial aviation and air travel is replete with stories of female harm, damage, and exclusion. Industry press releases from 2024 containing photographs of small groups of almost exclusively white men proudly inspecting the latest digital electrification, alternative fuelling systems, and hydrogen aviation-related 'techno-fixes' arguably do little to assuage potential concerns that aviation's negative environmental externality effects – which were largely created by men (female aircraft designers and engineers rivalling William Boeing and Frank Whittle are conspicuous by their absence), for (business) men (on which see Beaverstock et al., 2009) – can apparently also be fixed by men with no visible input from women. This is potentially problematic on a number of fronts.

The majority of the decarbonisation solutions that are currently being proposed – not just for aviation but also other transport modes – are largely based on technological innovations, some nascent, some more advanced, and commercially scalable. These range from the electrification of heavy trucks, railways (see Chapter 9) and private cars (Chapter 5) through to alternative aviation fuels and increased automation leading to enhanced efficiency. The dominant decarbonisation narrative within transport is thus one of technological determinisms in which technological innovations, perhaps implemented alongside ill-defined references to 'behavioural change' (or, just occasionally, 'demand management'), will give global society a metaphorical 'get out of jail' card while allowing its wealthier members to maintain their current levels of mobility and consumption with little inconvenience or additional cost.

The development of these new technologies, fuels, and digital infrastructures relies on the availability of finance for research and development, and individuals with the skills, training, and experience to upscale innovations and bring them to market. Although the growth of the wider 'green economy' and the focus on 'green jobs' to deliver low-carbon transport have been largely welcomed, it is important to recognise the inherently gendered nature of the new jobs that are being created. To date, many of the new jobs that have been created have been taken by men and the low-carbon technologies that are being brought to market are often designed by male-dominated and male-led teams. This reinforces a developing narrative that 'green jobs' in transport are not 'women's work' (cf. Jaegler et al., 2023, for a discussion of how conventional transport and logistics jobs were similarly historically deemed not women's work). Such concerns are not merely anecdotal, but they are borne out in contemporary employment figures. In the United States, it is estimated that only a third of green jobs are held by women (Gammon, 2023), while Sqalli et al. (2021) estimate that while the green economy will create 67 million new jobs by 2030, women will hold only 25% of them. Given these jobs are generally well paid, the gender disparity within the sector will perpetuate and entrench gender inequalities in pay and opportunities but also prevent employers from benefiting from the skills and knowledge arising from a more diverse workforce.

Some of this gender imbalance results from what is arguably a self-fulfilling consequence of pre-existing gender inequalities in further and higher education. Indeed, back in 2015, the International Labour Organization (ILO) (2015) expressed concerns that women's access to green jobs may be limited due to inadequate training and experience in subjects including engineering and computer science. Data from the United Kingdom's Higher Education Statistics Authority (HESA) reveal that only 27% of UK university students graduating with a degree in a STEM subject (science, technology, engineering, and mathematics) in 2022 identified as female or non-binary, while the proportion in some specific subjects was as low as 21% (STEM Women, 2023). Not surprisingly, this under-representation continues into the workforce. Women currently comprise around a quarter (26%) of the UK's STEM workforce, and it is estimated that equal representation in STEM careers will not occur until 2070 (STEM Women, 2023). Given the growing development and emphasis on introducing 'green' low-carbon technologies into the transport sector, and the relative lack of women working in these spheres, the potential implications for women working in transport and women using transport need to be considered, and it is to these issues that this chapter now turns.

WOMEN, WORK, AND TRANSPORT

The heading of this section is taken from the eponymous 2022 publication which formed volume 16 of this *Transport and Sustainability* series (Wright et al., 2022). The 23 chapters in the volume explored the nature of women's work in the transport sector and exposed some of the particular gender-related challenges and experiences of women working in a range of sectors and roles from seafarers to drivers of heavy-goods vehicles and the role of trades unions. In every case, the chapters exposed the fact that women were not only under-represented in the workforce but also exposed to situations specific to their gender. This ranged from dangers from physical and sexual assault to inappropriate and derogatory comments from colleagues, passengers, and the wider public. None of the chapters explicitly addressed women, work, and the gendered dimensions of transport decarbonisation, so this section adds an important dimension to that volume.

Although the absolute numbers and relative percentage of women working at all levels in different transport modes remain considerably lower than men at around 17% of the total global workforce (Ng & Acker, 2020), research has shown that diversity improves well-being and shareholder returns and means diverse businesses outperform their competitors on a range of key performance, reputational, and financial metrics (Chang, 2019). There are thus compelling business reasons, as well as moral and ethical imperatives, to improve gender diversity in the transport workforce. In recognition that the transition to a net zero economy (which includes transport) will not be gender neutral, the UK government and major international organisations have called for a 'gender-just' transition which considers the impact of net zero on all workers (male, female, those with other gender identities, and those working in the formal and informal sectors alike).

Historical examples have shown that growing automation in transport may result in technological displacement and the loss of previously skilled jobs. This is not a phenomenon solely impacting women, of course, and nor is it a modern one. From the mid-1970s onwards, the flight engineer's position (the 'third man' on the flightdeck) began to be replaced by computerised flight management systems, while on the rails, automated driverless transit systems, including London's Docklands Light Railway (DLR) which began operation in 1987, became more commonplace. The ability to achieve higher levels of economic output for lower levels of labour input became manifest with the introduction, in 1974, of Airbus' new A300-600 aircraft into revenue service. The A300 was the first wide-bodied commercial aircraft to be operated by a two-person crew. The fourth (known as the 400) series of Boeing's B747 'jumbo jet', which first flew in 1988, could similarly be operated by just two pilots, delivering significant savings on labour costs by effectively eliminating a skilled position. There is the potential for the next generation of small electrically powered aircraft to be wholly automated and operated with no crew. While the developers of such technology emphasise the environmental benefits of electrification, there are also gendered aspects both of developing and using transport systems which need to be considered, and these will be addressed in the following section.

As well as the potential for the displacement of highly skilled positions, there is also the potential for automated systems to replace conventional manual (often customer-facing) positions. In the case of transport, this includes railway ticket offices and manual check-in at airports, processes which are both increasingly being replaced by digital self-service systems. There is a clear gendered dimension to this. On the one hand, as women tend to disproportionately occupy more precarious and insecure positions, they may be disproportionately impacted by new technologies and in a position of relative powerlessness to do anything about it. On the other hand, displacing women and other workers away from front-line public-facing positions at airport check-in desks or on gate lines at railway stations, for example, may actually protect them from the verbal aggression and physical violence they experience (see Van Barneveld & Jowett, 2005).

This issue is arguably particularly acute in the global supply chains of multinational companies, especially those such as the agriculture and garment manufacturing sectors which rely on significant numbers of women. These women often work in poorly paid and/or informal and insecure employment with limited or no employment rights. While often considered to be an issue that is particularly acute in the global south, countries of the global north are not immune from women working in poor conditions. During the COVID-19 pandemic, reports of abuse and poor working conditions experienced by women workers in clothing factories in Leicester, UK, made headline news (Stewart, 2024). Although working in a major city in the global north, garment workers in the city who were supplying products to major clothing brands were not immune to labour abuses, including sudden dismissal, illegally low pay, and the withholding of wages. While sourcing products from factories close to their markets reduced the carbon intensity of transporting the finished goods to consumers, the practice was not without its human costs, which were largely experienced by women.

For Jenkins et al. (2023, p. 4), 'transitioning to low carbon supply chains presents businesses with an opportunity to simultaneously tackle the double injustice of climate change and gender inequality, leading to a triple win for people, economies and the planet'. This aspiration can, however, only be realised if the supply chains involved are truly committed to tackling these injustices. An important feature of any supply chain is, of course, the transport and logistics which are required to transport raw materials to the manufacturing centres, the finished goods to consumers, and (increasingly) sending goods returned by (online) buyers back to distribution centres for resale (see Chapter 12). The global logistics sector is thus also an important area for transport decarbonisation and gender consideration, and as with the supply chain, it has an important geographic and gendered dimension.

There is considerable evidence that countries of the global north engage in practices of 'technological dumping' in which older (and often more polluting and less-efficient vehicles) are transferred to the global south as older vehicles are replaced by newer models. This practice has a range of social, economic, and environmental implications, including concentrating pollution in new areas and creating a dependency on companies in the global north to supply replacement parts. Within the logistics sector, older more polluting and less-efficient diesel railway locomotives have been exported to parts of sub-Saharan Africa (see Bullock, 2009), obsolete ships are broken up in sometimes dangerous conditions in countries of the global south (Mannan et al., 2024), while older aircraft may be relocated to markets in the global south that do not have the capital required to purchase new equipment (see de Oliveira Junior et al., 2021).

Although there is, of course, an environmental case for continuing to use working machinery that already exists rather than manufacturing new products, there is a need to be aware of the indirect consequences (for all genders) of transferring older equipment and vehicles to the global south with all the attendant pollution, health, and safety issues this entails. While 'donating' old railway locomotives to countries in Africa, for example, might appear to be an act of generosity, Bullock (2009) reports on the challenges and inefficiencies imposed on countries which try to integrate multiple pieces of (perhaps unreliable) equipment from various manufacturers into their networks while lacking the engineering expertise and equipment to support them. Such interventions have both direct and indirect consequences for transport workers and users in these locations which need to be recognised. There is also a risk that the financial cost of developing and operating low-carbon transport technologies will mean that they are only operated in the global north, and so countries in the global south do not benefit as readily from them. The issue of the financial cost of transport decarbonisation leads into the third section. New technology is usually more expensive than the conventional technology it seeks to replace. This is likely to affect not only poorer nations but also women in all nations who, on average, earn less than men and have less disposable income than men and so may be priced out of using it and/or prevented from being early owners and adopters of low-carbon transport vehicles.

WOMEN AS USERS OF TRANSPORT SYSTEMS

This chapter has already exposed potential inflection points around the gendered dimensions of transport decarbonisation technologies and the importance of women being involved at every level of transport policy, planning, and provision. Arguably nowhere is this more manifest than in the current debates surrounding transport electrification and the gender(ed) implications for consumers of the growth of electric vehicles (EVs).

Private Transport

EVs are widely promoted for their air quality benefits and their elimination of tail pipe emissions (Chapter 5). Leaving aside challenging questions surrounding the mining and refining of the rare earth metals needed for the batteries and the roll out of charging stations, there is also an issue of who is using them. In November 2021, 76% of EV drivers in the United Kingdom were men (Statista, 2021), and men are more likely to purchase, own, and drive an EV than women. The equity impacts of the EV transition, when it comes to financial incentives, purchasing power, and access to EVs (whether as privately owned vehicles or as part of a corporate fleet) are thus not gender neutral (see Israel et al., 2023). Possible reasons for the gender disparity in EV uptake disparity were explored in an industry report commissioned by *AutoTrader* (2023), a major UK online vehicle sales marketplace. In the introduction to the report, which was titled *No Driver Left Behind. Women and the Journey to Electric,* Erin Baker, the Editorial Director, noted that:

> Men have historically designed, engineered, manufactured, marketed, advertised, sold, bought, driven, written and read about cars. Women have been excluded, ignored, patronised and mansplained to by brands, marketing and the industry, almost to the point of total alienation. All of which is a strange approach to engaging half your market ....

The report went on to reveal that while women were more concerned about EV range, cost, and safety and were less interested in vehicle technology and eco 'virtue signalling' than men, vehicle advertisements largely emphasised the cars' technological capabilities and performance.

Nor is the gendered nature of driving limited to EVs. Research has shown that women typically drive smaller and lighter vehicles than men which means they are more likely to be seriously injured in the event of an accident (Sweet, 2021; Welsh & Lenard, 2001). Debates over the desirability of using 'female' crash test dummies, which take into account the physiological differences between women and men, rather than using scaled-down male versions (see Elsesser, 2024) have exposed long-standing gender discrimination in the automotive sector. Whether the testing of vehicles using scaled-down male dummies in crash tests rather than dummies that accurately represented female bodies was a deliberate policy to overlook or downgrade the safety of women in cars is a moot point and beyond the scope of the present chapter, but the issue does pose important questions for how future low-carbon transport vehicles are designed, tested, and approved.

The issue of personal safety when driving extends beyond the avoidance of physical harm in the event of an accident, to issues of personal safety on the road. When driving, women report feeling intimidated by other road users, are more likely than men to drive with locked doors, and less likely to drive at night due to complex intersections of risk perception, confidence, fear of road rage, carjacking or assault, and feelings of responsibility to others, not least as women often transport children (Lyons, 2004). Others express relief at being able to travel alone without the attendant fear and anxiety of travelling with strangers on public transport, especially after dark and on unfamiliar routes. Given that much of the emphasis on decarbonising local transport and effecting mode shift to active travel and public transport modes will affect women, there is a need for greater awareness of the gender implications of such policies.

Public Transport

Women use public transport for a variety of reasons – they have no other choice, they use it to supplement car journeys, they prefer it, or they consciously choose to avoid it as far as possible (see Ng & Acker, 2018). The reasons are almost as diverse as the people themselves and use of public transport differs over the life course. Teenagers and younger adults are more likely to use buses than middle-aged women on account of their being too young to hold a driving licence or seeking independence but unable to afford the costs of learning to drive, purchasing, maintaining, running, and insuring a car. Middle-aged women may be less inclined to use buses as major life events, including new careers and changing domestic responsibilities, incentivise them to purchase a car (Oakil, 2016). Older adults are more likely to use buses following driving cessation on grounds of confidence, illness, or frailty (see Chapter 17).

Despite interventions designed to improve safety in the public realm and on vehicles through the deployment of CCTV and security staff, research has shown that a high proportion of women face frightening situations in their everyday mobility which constrains their mobility and directly influences their mode choice (Kacharo et al., 2022; Stark & Meschik, 2018). Multiple studies have shown women experience feelings of anxiety and fear while waiting for public transport (Chowdhury, 2022; Chowdhury & van Wee, 2020) or while travelling in empty carriages, and that concerns for personal safety are particularly acute after dark. Indeed, fear-induced nocturnal immobility (which is, of course, different for different age groups) means women typically do not go out as much at night, arrive home earlier, and/or purposefully avoid certain areas, stations, or routes (Zhang et al., 2022).

The absence of staff on automated transit systems is another area of anxiety for some women. This has implications for the proposed introduction of new forms of electric short-range urban air mobility vehicles which are being designed to operate without crew. Nascent research into the public perceptions of such technologies indicates a reluctance on behalf of some individuals to commit to travelling in a vehicle with no human operator and no means of summoning help in the event of an undesirable situation (Jones et al., 2024). Conversely, it could also be argued that autonomous vehicles may actually increase the mobility of women and vulnerable groups who prefer to travel in private vehicles rather than use public transport. Consequently, the outcome in terms of emissions and decarbonisation is unknown.

Designing vehicles that enable women to travel amid a critical mass of other travellers (thus promoting their personal safety) is something that can also assist with other forms of inclusion, particularly for users with disabilities. New trains are often more disability accessible than the older stock they replace, so helping both the environment and inclusivity. The 'new' S stock Metropolitan, District, Circle and Hammersmith, and City Line trains in London, which began to be introduced in 2013, are not only air conditioned but also permit step free access and are walk through. This means passengers can access the whole length of the train once inside it and move to different locations within it, without having to alight at a station to swap carriages. Such design, of course, promotes social mixing and prevents segregation and is the exact opposite of the policy of operating 'women-only' carriages on some subways and urban railways (cf. Tillous, 2019) or women-only taxis driven by female drivers. As the World Bank (2024, unpaginated) notes,

> Women and men have different mobility patterns and experience different risks when using public transport … This phenomenon is directly linked to the persistence of gender roles and gender stereotypes, which leads to structural inequalities in how women and men use their time, make decisions within the household, and distribute everyday responsibilities.

CONCLUSION

Achieving rapid decarbonisation in transport is urgent and is an important area of development and academic inquiry. However, a focus on transport decarbonisation on its own is arguably insufficient. This chapter has posited that transport decarbonisation strategies need to incorporate a gender perspective and involve a more diverse workforce to ensure that future transport systems are socially and environmentally equitable and that no gender is disproportionately privileged or disadvantaged. This will require wide-ranging conversations around all elements of the transport service and transport delivery, from multinational supply chains to local transport provision. Undoubtedly, more research will be needed to develop the instruments and expertise that will be needed to collect more nuanced gender-disaggregated and gender sensitive data that will enable academics and policymakers alike to analyse the gendered nature of transport decarbonisation and examine how it affects different communities around the world. There will also be a need to introduce (perhaps mandatory) gender-aware reporting to show how companies and public organisations are incorporating gender awareness into their transport planning and provision.

The importance of gender in transport decarbonisation is an issue which, until recently, has been largely overlooked in the academic literature. The fact that women are still poorly represented in the transport workforce, marginalised in the pursuit of 'green jobs' and forced to use a transport system that is not always aware of or appropriate to their needs is something that requires urgent attention. By reviewing the current research landscape into gender, decarbonisation, and transport, this chapter has shown the importance of politicians, policymakers, practitioners, and scholars being gender aware. Increased digitisation, automation, and investment in low(er)-carbon transport have the potential to transform mobility patterns and practices worldwide. A transport system that is not designed for and with everyone (irrespective of age, gender, race, nationality, or disability) will not achieve its decarbonisation targets, and so there is both moral and practical imperative to ensure that the gender agenda is forefront in the transition to net zero transport. Transport is at a pivotal moment. Its future direction will have profound implications for individuals, society, and the planet.

REFERENCES

Adey, P., Lin, W., Barry, K., Harris, T., Frétigny, J.-B., & Budd, L. (2024). Now boarding: Towards new geographies of aeromobility. *Progress in Human Geography*, *48*(6), 716–736.

Amaugo, A. (2022). Examining the culture of women in aviation leadership: A case of UK airports. In T. Wright, L. Budd, & S. Ison (Eds.), *Women, work and transport* (pp. 189–206). Emerald.

AutoTrader. (2023). *No driver left behind. Women and the journey to electric*. https://www.nodriverleft-behind.co.uk/

Beaverstock, J. V., Derudder, B., Faulconbridge, J. R., & Witlox, F. (2009). International business travel: Some explorations. *Geografiska Annaler: Series B, Human Geography*, *91*(3), 193–202.

Borgato, S., Maffii, S., Malgieri, P., & Chiffi, C. (2021). Women and gender-related aspects. In T. Kuttler & M. Moraglio (Eds.), *Re-thinking mobility poverty* (pp. 113–123). Routledge.

Bullock, R. (2009). *Off track: Sub-Saharan African railways*. Africa Infrastructure Diagnostic Paper No. 17. The International Bank for Reconstruction and Development/The World Bank. https://ppp.worldbank.org/public-private-partnership/sites/ppp.worldbank.org/files/documents/Africa_Offtrac%20-%20SubSaharan%20African%20Railways_EN.pdf

Carver, A., & Veitch, J. (2020). Perceptions and patronage of public transport – Are women different from men? *Journal of Transport & Health*, *19*, 100955.

Ceccato, V., & Loukaitou-Sideris, A. (2022). Fear of sexual harassment and its impact on safety perceptions in transit environments: A global perspective. *Violence Against Women*, *28*(1), 26–48.

Chang, J.-H. (2019). Unpacking the business case for gender diversity. In C. Kuptsch & É. Charest (Eds.), *The future of diversity* (pp. 15–29). ILO.

Chowdhury, R. (2022). Sexual assault on public transport: Crowds, nation, and violence in the urban commons. *Social & Cultural Geography*, *24*(7), 1087–1103.

Chowdhury, S., & van Wee, B. (2020). Examining women's perception of safety during waiting times at public transport terminals. *Transport Policy*, *94*, 102–108.

D'Agostino, C., Piva, E., Pucci, P., & Rossi, C. (2024). A systematic literature review on women's daily mobility in the Global North. *Transport Reviews*, *44*(5), 1016–1044.

De Oliveira Junior, F., Fernandes, E., Bahiense, L., & Grandi, C. (2021). A practical approach to support end-of-life commercial aircraft parking, market relocation, retirement and decommissioning strategic decisions. *International Journal of Production Research*, *59*(17), 5144–5163.

Desrochers, J. E., Albert, G., Milfont, T. L., Kelly, B., & Arnocky, S. (2019). Does personality mediate the relationship between sex and environmentalism? *Personality and Individual Differences*, *147*, 204–213.

Elsesser, K. (2024, February 20). Will female crash dummies improve road safety for women? The answer is complicated. *Forbes*. https://www.forbes.com/sites/kimelsesser/2024/02/20/will-female-crash-dummies-improve-road-safety-for-women-the-answer-is-complicated/

Francis-Devine, B., & Hutton, G. (2024, March 4). *Women and the UK economy*. Research Briefing No. 6838. House of Commons Library. Retrieved September 13, 2024, from https://researchbriefings.files.parliament.uk/documents/SN06838/SN06838.pdf

Gammon, K. (2023, December 13). 'Quite a gap to close': Women 'vastly underrepresented' in green jobs sector. *The Guardian*. https://www.theguardian.com/us-news/2023/dec/13/green-energy-jobs-women-workforce-solar-wind

Hopkins, D., Gössling, S., Cohen, S., Hanna, P., & Higham, J. (2023). Aero mobilities and the fallacy of sustainable aviation. *Energy Research & Social Science*, *106*, 103319.

International Energy Agency. (2024). *Transport*. Retrieved September 9, 2024, from https://www.iea.org/energy-system/transport

International Labour Organization. (2015). *Gender equality and green jobs*. Policy Brief. ILO. Retrieved September 11, 2024, from wcms_360572.pdf

International Transport Forum. (2021). *ITF transport outlook 2021*. OECD Publishing.

International Transport Forum. (2022). *Gender equality and the role of women in decarbonising transport ITF with the FIA Foundation*. OECD. Retrieved September 11, 2024, from https://www.itf-oecd.org/sites/default/files/docs/gender-equality-women-decarbonising-transport.pdf

Israel, F., Ettema, D., & van Lierop, D. (2023). Mechanisms with equity implications for the (non-) adoption of electric mobility in the early stage of the energy transition. *Transport Reviews*, *44*(3), 659–683.

Jaegler, A., Ruel, S., Kafa, N., & Budd, L. (2023). Gender diversity in logistics. In J. Monios, L. Budd, & S. Ison (Eds.), *The Routledge handbook of urban logistics* (pp. 387–394). Emerald.

Jenkins, O., Kanani, P., & Liddiard, C. (2023, December). *Supply chain decarbonisation with a gender lens. Practical guidance for multinational companies planning Scope 3 emissions reductions*. WOW Programme, UK Foreign and Commonwealth Office.

Jones, P., Budd, L., & Ison, S. (2024, October). *Embedding disabled passenger needs into the UK's advanced air mobility ecosystem*. Final Report. De Montfort University, Leicester. https://hdl.handle.net/2086/24364

Kacharo, D. K., Teshome, E., & Woltamo, T. (2022). Safety and security of women and girls in public transport. *Urban, Planning and Transport Research*, *10*(1), 1–19.

Kronsell, A., Rosqvist, L. S., & Hiselius, L. W. (2016). Achieving climate objectives in transport policy by including women and challenging gender norms – The Swedish case. *International Journal of Sustainable Transportation*, *10*(8), 703–711.

Lyons, G. (2004). Transport and society. *Transport Reviews*, *4*, 485–509.

Mannan, B., Rizvi, M. J., & Dai, Y. (2024). Ship recycling in developing economies of South Asia: Changing liability to a commodity. *Green Technologies and Sustainability*, *2*(2), 100064.

McCarthy, F., Budd, L., & Ison, S. (2015). Gender on the flightdeck: Experiences of women commercial airline pilots in the UK. *Journal of Air Transport Management*, *47*, 32–38.

McLaughlin, H., & Fearon, C. (2022). Closing the gender gap in the maritime industry: A career-decision approach. In T. Wright, L. Budd, & S. Ison (Eds.), *Women, work and transport* (pp. 265–278). Emerald.

Ng, W. S., & Acker, A. (2018). *Understanding urban travel behaviour by gender for efficient and equitable transport policies*. International Transport Forum Discussion Paper No. 2018-01. OECD Publishing, Paris.

Ng, W. S., & Acker, A. (2020). *The gender dimension of the transport workforce* [International Transport Forum Discussion Paper No. 2020/11]. OECD Publishing, Paris.

Oakil, A. T. M. (2016). Securing or sacrificing access to a car: Gender difference in the effects of life events. *Travel Behaviour and Society*, *3*, 1–7.

Polk, M. (2003). Are women potentially more accommodating than men to a sustainable transport system in Sweden? *Transportation Research Part D*, *8*, 75–95.

Rosqvist, L. (2019). Gendered perspectives on integrating gender into transport planning. In C. Scholten & T. Joelsson (Eds.), *Integrating gender into transport planning* (pp. 69–87). Palgrave Macmillan.

Smith, W. E., Cohen, S., Kimbu, A. N., & de Jong, A. (2021). Reshaping gender in airline employment. *Annals of Tourism Research*, *89*, 103221.

Sqalli, Z., Unnikrishnan, S., Mejri, N., Dupoux, P., George, R., & Zrikem, Y. (2021, October 29). *Why climate action needs a gender focus*. Boston Consulting Group. Retrieved September 13, 2024, from https://www.bcg.com/publications/2021/climate-action-impact-on-gender-equality#:~:text=Factoring%20in%20the%20impact%20on%20women%E2%80%99s%20participation%20in%20green%20jobs

Stark, J., & Meschik, M. (2018). Women's everyday mobility: Frightening situations and their impacts on travel behaviour. *Transportation Research Part F: Traffic Psychology and Behaviour*, *54*, 311–323.

Statista. (2021). *Self-reported gender of electric vehicle drivers in the United Kingdom as of November 2021*. Retrieved September 11, 2024, from https://www.statista.com/statistics/1314989/uk-ev-drivers-by-gender/#:~:text=76%20percent%20of%20respondents%20driving,in%20the%20UK%20in%202021

STEM Women. (2023, August 30). *Women in STEM statistics: Progress and challenges.* Retrieved September 11, 2024, from https://www.stemwomen.com/women-in-stem-statistics-progress-and-challenges

Stewart, H. (2024, January 14). 'She agreed to work for £5 an hour': The desperate plight of many Leicester garment workers. *The Guardian.* Retrieved September 13, 2024, from https://www.theguardian.com/business/2024/jan/14/she-agreed-to-work-for-5-an-hour-the-desperate-plight-of-many-leicester-garment-workers

Sweet, K. (2021, February 12). Smaller cars may be why women are more likely to suffer serious crash injuries. *The Insurance Journal.* Retrieved September 11, 2024, from https://www.insurancejournal.com/news/national/2021/02/12/601173.htm

Tillous, M. (2019). Women, (railway) class and the state: An analysis of two controversies surrounding women-only metro carriages (Cairo – São Paulo). *Gender, Place & Culture, 27*(8), 1155–1175.

Van Barneveld, K., & Jowett, R. (2005). Violence, harassment, and bullying at work: How does the Australian rail industry compare and what can be done? *Journal of Public Transportation, 8*(3), 117–134.

Welsh, R., & Lenard, J. (2001, September 24–26). Male and female car drivers – Difference in collision and injury risks. In *Proceedings of the 45th annual AAAM conference*, Texas. https://repository.lboro.ac.uk/articles/online_resource/Male_and_female_car_drivers_-_difference_in_collision_and_injury_risks/9339611?file=16948286

World Bank. (2024, March 7). *Closing gender gaps in transport.* World Bank Group. Retrieved September 11, 2024, from https://www.worldbank.org/en/topic/transport/brief/closing-gender-gaps-in-transport

Wright, T., Budd, L., & Ison, S. (Eds.). (2022). *Women, work and transport* (Transport and Sustainability, Vol. 16). Emerald Publishing Limited.

Zhang, M., Zhao, P., & Tong, X. (2022). Constructing women's immobility: Fear of violence and women's constricted nocturnal travel behaviour. *Travel Behaviour and Society, 26*, 178–192.

CHAPTER 17

AGEING, TRANSPORT, AND DECARBONISATION

Charles Musselwhite

Aberystwyth University, UK

ABSTRACT

It is often assumed that older people must rely on private transport in order to meet their needs. Indeed, research has found that older people find it very difficult to give up driving, as doing so can result in isolation and negative health effects. Intergenerational justice nevertheless suggests older people should not be treated as any special case, and it is imperative that interventions aimed at reducing high carbon, private vehicle use also include older people. Studies have found that older people want to be part of the decarbonisation agenda, can feel guilty for their contribution to climate change, and want to make a difference and play a role in it themselves. However, the current older cohort in many countries consists of people who have driven the most of any generation, and targeted approaches that focus on social and affective aspects of reducing carbon-intensive mobility may be needed. Good communication, information, trust, and the use of demonstrators and champions can help older people move to more sustainable mobility and enable better carbon literacy, which in turn informs other more sustainable behaviours.

Keywords: Ageing; gerontology; transport; sustainability; health; climate change

Towards Transport Net Zero
Transport and Sustainability, Volume 20, 255–267

ISSN: 2044-9941/doi:10.1108/S2044-994120250000020017

INTRODUCTION

Populations across the world are ageing. In 2024, there were 750 million people aged over 65 across the globe, and predictions suggest that this is to rise by over double to 1.6 billion by 2050. Over the same period, the proportion of over 60s will rise from 10% to 16% of the population (United Nations, 2022). Older people across the globe are more active and more mobile than previous generations, and their use of transport is a growing contributor to climate change.

This chapter reviews the key issues associated with ageing, transport, and decarbonisation. It begins by addressing the importance of mobility within an increasing ageing society, identifying that the car has become the default form of transport for older people across the globe and the difficulties in reducing or ceasing driving for that group. It then examines how older people are more at risk of changes to the environment as a result of climate change. This chapter then moves to look at the attitudes, acceptability, and motivations of older people in relation to sustainability and net zero in a wider transport context. It looks at intergenerational justice and the importance of bringing older and younger people together before concluding with some policy interventions and the need for practical and emotional support around using more sustainable forms of transport in later life. Throughout this chapter, older people are defined as those over the age of 70 unless otherwise specified.

HEALTHY AGEING AND TRANSPORT USE

The World Health Organization (WHO, 2020, unpaginated) has outlined the need for all societies to move towards healthy ageing, defined as ‘the process of developing and maintaining the functional ability that enables wellbeing in older age’. It goes on to note that ‘Healthy ageing is about creating the environments and opportunities that enable people to be and do what they value throughout their lives’, and in this context, transport is critical in supporting older adults’ mobility, health, and well-being needs (Mackett, 2018; Musselwhite & Haddad, 2010, 2018).

The nature of travel changes in later life. Older people are less likely to be in education or working, and this, coupled with health changes that impact on travel, mean they are responsible for fewer miles per year than the population as a whole. That said, in many countries, there has been an increase in the number of miles travelled by older people over the past decade or so (Mackett, 2018). Older people in middle- and high-income countries that are ‘hypermobile’ – i.e., they have very high levels of mobility – are more than ever likely to be driving to meet their everyday needs (Musselwhite & Curl, 2018; Musselwhite & Murray, 2024; Musselwhite & Scott, 2019). In the United Kingdom, for example, between 2002 and 2022, the distance travelled per person aged over 70 has increased by 22%, but driving has increased by 50% (Department for Transport (DfT), 2023). And whereas in 2002, 45% of over 70s held a driving licence, 20 years later in 2022, this had risen to 77% (DfT, 2023).

In the United Kingdom, as in many high-income countries, older people today have driven more across the life course than any other older generation (DfT, 2023). They have formed habits around driving for satisfying their mobility and social needs, and this is behaviour that becomes difficult to break. In studies where disruption has occurred at a societal level – e.g., COVID-19 and associated travel restrictions forbidding all but essential journeys (Musselwhite et al., 2020, 2021) – or at the individual level – e.g., breaking a wrist and having it in plaster restricting the ability to drive (Musselwhite et al., 2016) – people find it difficult to stop driving, and some even continue to do so when really they should have stopped. The difficulty in breaking habitual and normal behaviour is even more marked when the requirement to cease driving is viewed as more permanent due to changes in health associated with ageing, such as deteriorating cognition or worsening eyesight and hearing (Mitchell, 2018).

Giving up driving is typically associated with negative health outcomes, a perceived loss of independence and freedom, and increased isolation and loneliness (Musselwhite & Haddad, 2010, 2018). Hence, older people are often reluctant to voluntarily cease driving, citing concerns over difficulty in accessing what they need – healthcare appointments, services, shops, meetings with friends and family, and so on (Mackett, 2018). As such, the need to shift from the use of private vehicles among older people to help reach net zero requires increased understanding of issues with transportation and ageing in the context of these wider negative associations with driving cessation.

At the same time, climate change is a significant threat to healthy ageing, with older people especially susceptible to the physical and mental health impacts of climate change (Park et al., 2020). Older people are more likely to pass away following extreme weather and in any subsequent evacuation (Aida et al., 2017; Åström et al., 2011; McClelland et al., 2017; Willoughby et al., 2017). Heating of the planet has direct negative health impacts on cardiovascular health and can cause kidney damage and disease (Borg et al., 2019; Kenney et al., 2014). Recent research is also beginning to identify relationships between exposure to air pollution and dementia (Peters et al., 2019). Overlaying all of this, it is acknowledged that those most vulnerable older people with chronic medical conditions and those living in care homes or lacking resources are most likely to suffer negative effects of climate change (e.g., Finlay et al., 2012; United Nations, 2020).

Despite this, older people across the globe have been neglected from inputting into research and policy around the impact of climate change on health and debates about climate change mitigation. The WHO (2021) notes that although older people make up a significant proportion of people affected by emergencies related to climate change, funding for projects to investigate their needs remains low. Moreover, as the next section of this chapter identifies, there is certainly scope for older people to be more involved in debates about climate change and how these relate to what they feel are the unique needs of their age group.

MOTIVATIONS TO MOVE TOWARDS DECARBONISATION AND NET ZERO

Motivations differ within populations and older people represent a less homogeneous group in terms of attitudes, behaviour, health, and well-being than is often acknowledged (Ayalon & Senjooti, 2023). On the whole, older people have different energy-use profiles than younger adults (Morgan et al., 2024). For example, in middle- and high-income countries, they are more likely to spend time at home, require additional heating and lighting than younger people (Brounen et al., 2012; Morgan et al., 2024), and, as we have seen, are less likely to use transport and hence travel lower miles; in travelling fewer miles, they are less likely to use trains and drive as much as younger adults (Musselwhite & Murray, 2024).

As their energy use is different to the population as a whole, it is perhaps unsurprising that they are likely to have different attitudes to supporting means to change behaviour and engage in climate mitigation. Across the world, older people are somewhat less likely to believe that climate change is a result of human action and that it is something that requires direct human intervention and behaviour change to reduce or stop the effects (Milfont et al., 2021). They are more likely to believe that 'science will save us' (Moody, 2014, 2017) and less likely to get involved in climate change activism (Ayalon et al., 2022). These barriers potentially make older people's engagement in the net zero debate and climate mitigation strategies lower than other age groups.

It is also true that older people tend to be more likely to favour what is good for society compared to younger adults, a concept known as prosocial value orientation (see Rushton, 2004). In short, what older people see as 'what is good for society' makes a difference to how they may act. Older people tend, for example, to be influenced heavily by social norms and want to support family, friends, and communities close to them to deliver what they need. In transport terms, however, this often translates to supporting a 'business as usual', i.e., car orientated, position. This is especially the case in a society dominated by 'institutional car-ism' (Musselwhite, 2021) – the sometimes unconscious bias of placing the car higher in priority throughout society, including in policies and practices – or 'motor normativity' (Walker et al., 2023) – where people are afflicted by a cultural blindness to anything negative around the private vehicle.

Interestingly, though, it is not the car per se that holds special privilege in meeting older people's needs, but rather the mobility that is linked to health and well-being. And crucially, this doesn't have to be provided by cars. Musselwhite (2018) found that older people who had always used the bus use the same language about the importance of the freedom, independence, convenience, and connectivity provided by buses, as drivers do about cars. A well-designed transport system that includes accessible public transport and active travel provision can support older people's health and well-being and their transport needs as much as cars and would be important in reducing private vehicle use as part of a broader move towards net zero (Johnson et al., 2017; Musselwhite, 2018).

Many countries have adopted legislation and funding to create accessible public transport. In the United Kingdom, for example, the *Access for All* programme delivers funding to allow equal access for everyone, regardless of their needs, to use the railway and includes station and rolling stock guidance. Nevertheless, where older people have come to perceive cars as being the main, or indeed only, source of comfortable, convenient, cost-effective mobility, policymakers have to take this into account when developing net zero solutions because taking away vehicles, or changing the conditions of their use, can be perceived by drivers as curtailing people's freedom and independence (Musselwhite & Haddad, 2010, 2018; Watkins & Musselwhite, 2024).[1] As Parkhurst et al. (2014) note, understanding older people's transport needs reveals that policy needs to examine mobility in terms of accessibility, motivations, norms, and experience, rather than in economic terms of time and distance. As such, the provision of net zero interventions and policy for older people needs to be cognisant of the wider social or cultural context in which practical and infrastructural elements of mobility are delivered.

ACCEPTABILITY OF TRAVEL BEHAVIOUR CHANGE AND NET ZERO INTERVENTION

Researchers examining best practice in working towards net zero for an ageing population often neglect transport solutions. Instead, they focus on housing (e.g., using sustainable materials and energy-efficient technology for health and cooling – Morgan et al., 2024; WHO, 2012) or general urban and city planning, including green and blue spaces and biodiversity (Lee et al., 2016). Addressing noise and air pollution is also noted, as are better links to public spaces, services, and shops, but these aspects are more akin to urban design rather than transport per se (Hartig et al., 2014; Lee et al., 2016; Nowak et al., 2006).

One approach to reducing private vehicle use is road pricing. In a case study in Bristol, UK, Nikitas et al. (2011) found older people to be more negative towards the intervention than younger adults. This was in part because of anxieties about the financial implications of giving up their vehicle and using other modes of transport but also because of worries about the impracticality of changing modes when living outside of potential congestion charging cordons. The impracticality of other modes and the effort needed to switch mode from private vehicles are issues cited by many older people when considering or planning to move to alternative forms of transport, especially for those with increasingly poor health (Musselwhite & Haddad, 2010, 2018). Accessibility to public transport continues to be a key issue, requiring level boarding, space to sit down, and usable toilets, either onboard or at the destination (Musselwhite & Attard, 2021; Musselwhite & Roberts, 2021). In addition, the anticipation of future accessibility needs is enough to keep older people wedded to their vehicles (Musselwhite & Haddad, 2018).

Financial issues associated with changing modes are often cited where older people have retired and have lower income and assets than when they were working (Musselwhite & Haddad, 2018). The sunk cost of an owned vehicle tends to mean that only the direct costs of using it are considered in relation to potential alternatives. For example, older people might take into account only the costs of fuel and parking when making an immediate decision about whether to use the car or take a train, identifying (often erroneously) driving their car as the cheaper mode of transport. While this is similar to the population as a whole, it is exacerbated because older people do not drive as many miles, and hence, there remains an underlying and proportionally greater relevance of their vehicle's sunk cost ('I've paid for this and therefore I'm going to use it!'). Equally, it is important to note that older people have heightened need for something termed the 'potential for travel' (Metz, 2000) and have the strong perception that a car provides security in case something happens and is needed to support a friend or relative in hospital (or in some cases simply to have a day out for leisure or well-being purposes) (Musselwhite & Haddad, 2010, 2018).

Encouraging walking or cycling among older people meets unique barriers. Older people often feel they lack the physical ability to walk or cycle (Jones et al., 2018; Musselwhite, 2021). In addition, in certain societies, societal norms discouraging walking and cycling affect the amount older people ride a bike or go by foot. But in countries where it is normal to walk and cycle well into later life, and there has been better investment in infrastructure, older people remain active in later life and walk or cycle more for journeys they would otherwise take by car. The amount of cycling varies hugely from country to country. In the United Kingdom, for example, only 1% of trips made by older people are by bicycle, whereas in Denmark, it is 15% and in Holland, 23%, and in low-to middle-income countries, it is around 40%–60% (Pucher & Buehler, 2012).

In low- to middle-income countries where the prevalence of walking and cycling is high among older people, roads in urban areas are shared with high numbers of vehicles without adequate infrastructure, and in rural areas, roads are often used by high-speed vehicles, and there are no specific areas to walk and cycle, making them dangerous (Porter et al., 2018). That said, walking can be dangerous in high-income countries too; in Australia, for example, older people are 20% of the population, but make up 37% of all pedestrian road traffic collisions. Musselwhite (2021) notes how unsuitable the walking environment can be in many places, citing the need for better dedicated walking space away from vehicles and cars that is properly maintained and free from obstructions such as parked cars and street furniture. In addition, poor crossing facilities that don't stop the traffic long enough or are sited in the wrong places hinder older people's ability to cross the road (Musselwhite, 2015).

Older people are increasing users of e-bikes in many high- and middle-income countries where more older people can afford them, not least because they require less physical effort compared to existing bicycles (MacArthur et al., 2014; Van Cauwenberg et al., 2019). Indeed, electric vehicles more generally are increasingly becoming popular with older people. An initial lag or scepticism in take-up among older people compared to younger or middle-aged adults is waning. In the United Kingdom at least, environmental reasons for taking-up electric vehicles are lower in priority compared to younger or middle-aged adults, with convenience and cost being higher up on the agenda for older people (Musselwhite, 2019).

Thinking towards a future use of mobility, evidence from a UK study shows older people are quite in favour of autonomous vehicles (Musselwhite, 2019), albeit in the same 'business as usual' sense as noted above, where autonomous vehicles would simply replace current car driving activities, rather than promoting alternatives. Musselwhite (2019) did, however, find high levels of support among older people for better communication that technological systems that might better integrate transport, for example, mobility-as-a-service style offerings. Going further and replacing literal mobility with imaginative or digital mobility is flagged as a distinct policy option by Parkhurst et al. (2014). Needs can be met without recourse to actually being physically mobile, and this is a key resource in the move to net zero. Potential options are E-shopping, talking or meeting over the internet, or using apps such as *Window to the Outside World* which brings 'the outside in' through facilities such as Google Earth and Google Streetview (Dowds & Masthoff, 2015; Reynolds et al., 2018). The COVID-19 pandemic has amplified older people's use of technology for such activities (Murray, 2021), but there remains the potential for users to miss the essence of mobility, of being co-present with their environments and within other people's worlds (Musselwhite, 2019).

INTERGENERATIONAL ISSUES

Climate change can reveal tensions between different generations. People experiencing issues of climate change can blame previous generations for not adapting their behaviour. It is sometimes documented that younger people dealing with climate change blame older people for contributing towards the climate problems. However, research would suggest such intergenerational conflict is at least partly unfounded (Frumkin et al., 2012; Thomas et al., 2024): some older people feel guilty for the impact of their actions on the planet and wish to make amends in later life, and the concept of leaving a legacy in terms of improving planetary health is also noted.

In any case, intergenerational work between young and old can have distinct value by revealing how different life used to be, and how past practices can be related to action that today could help address climate change. Younger people can learn from older people's experiences in the past with lower carbon and net zero practices such as fewer (or no) car or aeroplane journeys, and more walking and cycling (see Corner et al., 2015). Conversely, where older people do not know that things they now view as negative about life are due to, or are contributing towards, climate change, intergenerational work can be revelatory and encourage them to think seriously about how climate change has altered their lives. In a UK study to co-produce a comic book addressing climate change, Thomas et al. (2024) documented older people sharing stories of low-carbon practices and skills in their youth, including direct transport activity (e.g., walking and cycling to work and education) and activities indirectly related to transport (e.g., growing your own food and buying local produce). Young people would then map these to potential low- or zero-carbon solutions onto their own lives, leading to discussions around climate mitigation strategies, including transport behaviour change. Overall, Thomas et al. (2024) demonstrated the benefits of involving and engaging older people more in discussions on transport solutions to climate change.

POLICY INTERVENTIONS

Given that older people can hold positive views towards sustainability and net zero, and intergenerational work has revealed they often hold knowledge, skills, and practices associated with moving towards net zero, policymakers can actively help support them to move towards more sustainable practices in their travel and transport behaviour. Policies such as the United Kingdom's *Access for All*, for example, could be expanded. Inaccessible infrastructure is a known barrier to using public transport for older people in many places across the world, and improvements to encourage public transport use that would suit older people include level boarding, reducing the need to use stairs and escalators with level movement at interchange or use of lifts, bright lighting, legible signage and information, priority seating provision, and the provision of toilets at interchanges/on services (Johnson et al., 2017).

Some of this is already in existence, but even the best physical infrastructure can't prevent anxiety if other aspects of the system fail or perform poorly – the service is late or breaks down or the journey requires several connections. This can be somewhat overcome through good information provision, including real-time service updates, but older people often feel much safer getting their updates from members of staff. Staff can make or break services for older people, even if they are relatively inaccessible, with practical help and emotional support (Musselwhite & Attard, 2021; Musselwhite & Roberts, 2021). Older people often want more tacit information around a service that is not usually provided, for example, knowing whether it is busy, where to stow bags and other items of luggage, and whether they will get a chance to purchase food. Hence, policy must move beyond simple infrastructure provision and work with older people to identify key aspects of what really makes a journey important for them.

Active travel can also require similar legislative support. Older people need dedicated infrastructure that is well kept, provided with crossing points but largely away from vehicles, free from clutter and safe to use, with regular spaces for sitting or provision of toilets. Legislation about minimal provision would help older people walk and cycle more. Again, social aspects are important as they are with a move to public transport; having support from other people to encourage walking and cycling can help create social norms around it for older people. Indeed, in helping older people drive less, more general support around giving up driving is needed. Older people are better equipped to give up driving when they plan early, familiarise themselves with alternatives, and have practical and emotional support from family and friends (Murray, 2021; Musselwhite & Murray, 2024). At the moment, in many countries, there is very little such support, and as such, it can be a very difficult process for people to go through.

There is also little support to help older people make the decision *when* to stop driving. One reason for this could be that countries that have more stringent testing in place with the aim to help older people give up driving – e.g., on-road testing, cognitive or medical screening – have no fewer collisions or injuries on the road than countries that have relatively relaxed measures (e.g., self-declaration at 70 years, and then every three years thereafter, as is the case in the United Kingdom; see Mitchell, 2018). This isn't to say testing is wrong – some type of testing or screening might well be useful – it's just that more research is needed to better determine what is needed in such tests to identify safe and dangerous drivers. Such tests do not come in isolation and have consequences in terms of economies and how society views older drivers (e.g., stereotyping them as dangerous).

In Denmark, a medical screening test was introduced, only to be taken away again as it made no difference to collision rates among older drivers while costing a lot of money. It was also labelled as 'ageist', emphasising to society that older people were dangerous drivers and potentially shifting blame from other dangerous drivers (Mitchell, 2018). A comparison of the effects of licencing for older people in Finland (regular medical check-ups after 70 years of age) and Sweden (no age-related controls) found no difference in vehicle crashes between the two countries, but Finland had a higher rate of fatalities among older pedestrians than Sweden because people walked more after having their licence taken away (Hakamies-Blomqvist et al., 1996).

In any case, support for moving to alternative modes of transport afterwards is largely found passively and presented online, although one intervention that has been successful that is group based and more dynamic, is the CarFreeMe initiative developed in Australia. This involves people working together to give up driving, normalising the process and navigating alternatives to the car (Scott et al., 2019) on the basis that providing practical and emotional support in social settings is a key part of the success of such an initiative.

CONCLUSION

Overall, older people need to be better understood within debates around climate change and net zero. Initially, the impacts of climate change on curtailing their behaviour and subsequent effects on health and well-being need further research, especially given the breadth of experience across populations of older people. The additional negative effects of climate change on older people's health and well-being also need recognising and responding to. As a result, there needs to be more involvement of older people in the development of net zero solutions. Good communication, information, trust, and the use of demonstrators and champions can help older people move to more sustainable and less carbon-intensive activity (see Morgan et al., 2024), and this will be the case in a mobility context. It is likely this provides better understanding of the key issues surrounding climate change, sustainability, and carbon literacy which informs other more sustainable behaviours as well.

People involved in delivering net zero policy and interventions can learn from older people's experiences of climate change and adaptation (Herman-Mercer et al., 2016), their use of sustainable behaviours across the life course (Biara et al., 2021), and their visions for sustainable futures (Leeson, 2015). Policy developed around helping older people become more sustainable with their travel choices needs to recognise the importance of providing accessible high-quality infrastructure to support public transport and active travel. Social support alongside this is also important. Finally, there is also a need to acknowledge the difficulty of giving up driving in hypermobile, car-dependent societies and the need for associated practical and emotional support.

Ultimately, it will not be possible to fully achieve healthy ageing without healthy, climate-resilient environments and a move towards net zero. Older people should be centrally involved in realising these things.

NOTE

1. Car manufacturers have sold their vehicles on this basis for well over the last century (Egan & Wright, 2000).

REFERENCES

Aida, J., Hikichi, H., Matsuyama, Y., Sato, Y., Tsuboya, T., Tabuchi, T., Koyama, S., Subramanian, S. V., Kondo, K., Osaka, K., & Kawachi, I. (2017). Risk of mortality during and after the 2011 Great East Japan Earthquake and Tsunami among older coastal residents. *Scientific Reports*, *7*(1), 1–11.

Åström, D. O., Bertil, F., & Joacim, R. (2011). Heat wave impact on morbidity and mortality in the elderly population: A review of recent studies. *Maturitas*, *69*(2), 99–105.

Ayalon, L., & Senjooti R. (2023). The perceived contribution of older people to climate change impact, mitigation, and adaptation: Measurement development and validation. *Innovation in Aging*, *7*(8), igad095.

Ayalon, L., Ulitsa, N., AboJabel, H., & Engdau, S. (2022). Older persons' perceptions concerning climate activism and pro-environmental behaviors: Results from a qualitative study of diverse population groups of older Israelis. *Sustainability*, *14*(24), 16366.

Biara, E., Egeru, A., Mensah, S., Salamula, J. B., & Kadigo, M. M. (2021). Socio-economic factors influencing Afzelia Africana Sm. use value and traditional knowledge in Uganda: Implications for sustainable management. *Environment, Development and Sustainability*, *23*, 2261–2278.

Borg, M., Nitschke, M., Williams, S., McDonald, S., Nairn, J., & Bi, P. (2019). Using the excess heat factor to indicate heatwave-related urinary disease: A case study in Adelaide, South Australia. *International Journal of Biometeorology*, *63*(4), 435–447.

Brounen, D., Kok, N., & Quigley J. M. (2012). Residential energy use and conservation: Economics and demographics. *European Economic Review*, *56*(5), 931–945. https://doi.org/10.1016/j.euroecorev.2012.02.007

Corner, A., Roberts, O., Chiari, S., Völler, S., Mayrhuber, E. S., Mandl, S., & Monson, K. (2015). How do young people engage with climate change? The role of knowledge, values, message framing, and trusted communicators. *Wiley Interdisciplinary Reviews: Climate Change*, *6*, 523–534. https://doi.org/10.1002/wcc.353

Department for Transport. (2023). *National travel survey*. Retrieved October 3, 2024, from https://www.gov.uk/government/statistics/national-travel-survey-2022

Dowds, G., & Masthoff, J. (2015). A virtual 'window to the outside world': Initial design and plans for evaluation. In S. Lawson & P. Dickinson (Eds.), *Proceedings of the 2015 British HCI conference (British HCI '15)* (pp. 265–266). Association for Computing Machinery. https://doi.org/10.1145/2783446.2783603

Egan, J., & Wright, C. (2000). De-marketing the car. *Transport Policy*, *7*(4), 287–294.

Finlay, S. E., Moffat, A., Gazzard, R., Baker, D., & Murray, V. (2012). Health impacts of wildfires. *PLOS Current Disasters*, *2*(4), e4f959951cce2c.

Frumkin, H., Fried, L., & Moody, R. (2012). Aging, climate change, and legacy thinking. *American Journal of Public Health*, *102*, 1434–1438.

Hakamies-Blomqvist, L., Johansson, K., & Lundberg, C. (1996). Medical screening of older drivers as a traffic safety measure: A comparative Finnish-Swedish evaluation study. *Journal of the American Geriatrics Society*, *44*(6), 650–653.

Hartig, T., Mitchell, R., Vries, D. S., & Frumkin, H. (2014). Nature and health. *Annual Review of Public Health*, *35*, 207–228.

Herman-Mercer, N. M., Matkin, E., Laituri, M. J., Toohey, R. C., Massey, M., Elder, K., Schuster, P. F., & Mutter, E. A. (2016). Changing times, changing stories: Generational differences in climate change perspectives from four remote indigenous communities in Subarctic Alaska. *Ecology and Society*, *21*(3), 535.

Johnson, R., Shaw, J., Berding, J., Gather, M., & Rebstock, M. (2017). European national government approaches to older people's transport system needs. *Transport Policy*, *59*, 17–27.

Jones, T., Chatterjee, K., Spencer, B., & Jones, H. (2018). Cycling beyond your sixties: The role of cycling in later life and how it can be supported and promoted. In C. Musselwhite (Ed.), *Transport, travel and later life* (Transport and Sustainability, Vol. 10, pp. 139–160). Emerald.

Kenney, W. L., Craighead, D. H., & Alexander, L. M. (2014). Heat waves, aging, and human cardiovascular health. *Medicine and Science in Sports and Exercise*, *46*(10), 1891–1899.

Lee, H., Mayer, H., & Chen, L. (2016). Contribution of trees and grasslands to the mitigation of human heat stress in a residential district of Freiburg, Southwest Germany. *Landscape Urban Planning*, *148*, 37–50.

Leeson, L. (2015). Case study: Engaging older people in creative thinking – The active energy project. In S. Clift & P. M. Camic (Eds.), *Oxford textbook of creative arts, health, and wellbeing: International perspectives on practice, policy and research* (pp. 245–250).

MacArthur, J., Dill, J., & Person, M. (2014). Electric bikes in North America: Results of an online survey. *Transportation Research Record*, *2468*, 123–130.

Mackett, R. (2018). Older people's travel and its relationship to their health and wellbeing. In C. B. A. Musselwhite (Ed.), *Transport, travel and later life* (Transport and Sustainability, Vol. 10, pp. 15–36). Emerald.

McClelland, E., Amlôt, R., Rogers, M. B., Rubin, G. J., Tesh, J., & Pearce, J. M. (2017). Psychological and physical impacts of extreme events on older adults: Implications for communications. *Disaster Medicine and Public Health Preparedness*, *11*(1), 127–134.

Mitchell, C. (2018). Are older people safe drivers on the roads, testing and training? In C. B. A. Musselwhite (Ed.), *Transport, travel and later life* (Transport and Sustainability, Vol. 10, pp. 37–64). Emerald.

Metz, D. H. (2000). Mobility of older people and their quality of life. *Transport Policy*, *7*(2), 149–152.

Milfont, T. L., Zubielevitch, E., Milojev, P., & Sibley, C. G. (2021). Ten-year panel data confirm generation gap but climate beliefs increase at similar rates across ages. *Nature Communications*, *12*(1), 4038.

Moody, H. R. (2014). Overcoming objections by elders to action on climate change. *The International Journal of Aging and Human Development*, *80*(1), 64–75.

Moody, R. (2017). Elders and climate change: No excuses. *Public Policy & Aging Report*, *27*(1), 22–26.

Morgan, D. J., Maddock, C., & Musselwhite, C. (2024). These are tenants not guinea pigs: Barriers and facilitators of retrofit in Wales, UK. *Energy Research and Social Science*, *111*, 103462.

Murray, A. (2021). *Winter stories*. Older People's Commissioner for Wales.

Musselwhite, C. B. A. (2015). Environment–person interactions enabling walking in later life. *Transport Planning & Technology*, *38*(1), 44–61.

Musselwhite, C. B. A. (2018). Mobility in later life and wellbeing. In M. Friman, D. Ettema, & L. Olsson (Eds.), *Quality of life and daily travel. Applying quality of life research (best practices)* (pp. 235–251). Springer.

Musselwhite, C. B. A. (2019). Older people's mobility, new transport technologies and user-centred innovation. In B. Müller & G. Meyer (Eds.), *Towards user-centric transport in Europe – Challenges, solutions and collaborations* (Lecture Notes Series, pp. 87–103). Springer.

Musselwhite, C. B. A. (2021). *Designing public space for an ageing population: Improving pedestrian mobility for older people*. Emerald.

Musselwhite, C. B. A., & Attard, M. (2021). Public transport use in later life. In C. Mulley, J. D. Nelson, & S. Ison (Eds.), *Routledge handbook of public transport* (pp. 393–404). Routledge.

Musselwhite, C. B. A., Avineri, E., & Susilo, Y. (2020). The coronavirus disease COVID-19 and implications for transport and health. *Journal of Transport and Health*, *16*, 100853.

Musselwhite, C. B. A., Avineri, E., & Susilo, Y. (2021). Restrictions on mobility due to the coronavirus Covid19: Threats and opportunities for transport and health. *Journal of Transport and Health*, *20*, 10142.

Musselwhite, C. B. A., Calcraft, M. J., Roberts, M., Fox, R., Swinkels, A., Turton, P., & Young, S. (2016). Breaking the habit: Does fracturing your wrist change your travel and driver behaviour? *Transportation Research Part F: Traffic Psychology and Behaviour*, *38*, 83–93.

Musselwhite C. B. A., & Curl, A. (2018). Geographical perspectives on transport and ageing. In A. Curl & C. Musselwhite (Eds.), *Geographies of transport and ageing* (pp. 3–24). Palgrave Macmillan.

Musselwhite, C. B. A., & Haddad, H. (2010). Mobility, accessibility and quality of later life. *Quality in Ageing and Older Adults*, *11*(1), 25–37.

Musselwhite, C. B. A., & Haddad, H. (2018). Older people's travel and mobility needs. A reflection of a hierarchical model 10 years on. *Quality in Ageing and Older Adults*, *19*(2), 87–105.

Musselwhite, C. B. A., & Murray, A. (2024). Travel behaviour of older people. In D. Potoglou & J. Spinney (Eds.), *Handbook of travel behaviour* (pp. 412–434). Edward Elgar Ltd.

Musselwhite, C. B. A., & Roberts, K. (2021). Accessibility and informational barriers to an age friendly railway. *Quality in Ageing and Older Adults*, *22*(2), 114–129.

Musselwhite, C. B. A., & Scott, T. (2019). Developing a model of mobility capital for an ageing population. *International Journal of. Environmental Research and Public Health*, *16*(18), 3327.

Nikitas, A., Avineri, E., & Parkhurst, G. (2011). Older people's attitudes to road charging: Are they distinctive and what are the implications for policy? *Transportation Planning and Technology*, *34*(1), 87–108.

Nowak, D. J., Crane, D. E., & Stevens, J. C. (2006) Air pollution removal by urban trees and shrubs in the United States. *Urban Forestry Urban Greening*, *4*, 115–123.

Park, C. E., Jeong, S., Harrington, L., Lee M. L., & Zheng, C. (2020). Population ageing determines changes in heat vulnerability to future warming. *Environment Research Letters*, *15*, 114043.

Parkhurst, G., Galvin, K., Musselwhite, C., Phillips, J., Shergold, I., & Todres L. (2014). Beyond transport: Understanding the role of mobilities in connecting rural elders in civic society. In C. Hennesey, R. Means, & V. Burholt (Eds.), *Countryside connections: Older people, community and place in rural Britain* (pp. 125–175). Policy Press.

Peters, R., Ee, N., Peters, J., Booth, A., Mudway, I., & Anstey, K. J. (2019). Air pollution and dementia: A systematic review. *Journal of Alzheimer's Disease*, *70*(s1), S145–S163.

Porter, G., Tewodros, A., & Gorman, M. (2018). Mobility, transport and older people's wellbeing in sub-Saharan Africa: Review and prospect. In A. Curl & C. Musselwhite (Eds.), *Geographies of transport and ageing* (pp.75–100). Palgrave Macmillan.

Pucher, J., & Buehler, R. (2012). *City cycling*. MIT Press.

Reynolds, L., Rodiek, S., Lininger, M., & McCulley, A. (2018). Can a virtual nature experience reduce anxiety and agitation in people with dementia? *Journal of Housing for the Elderly*, *32*(2), 176–193.

Rushton, J. P. (2004). Genetic and environmental contributions to pro-social attitudes: A twin study of social responsibility. *Proceedings of the Royal Society*, *271*, 25832585.

Scott, T. L., Liddle, J., Mitchell, G., Beattie, E., & Pachana, N. (2019). Implementation and evaluation of a driving cessation intervention to improve community mobility and wellbeing outcomes for people living with dementia: Study protocol of the, CarFreeMe, for people with dementia program. *BMC Geriatrics*, *19*, 66.

Thomas, M. J., Sorvala, L., Williams, A., Singleton, A., Maddock, C., Morgan, D., Murray, T., & Musselwhite, C. (2024). Co-creating a climate comic book: Reflections on using comics in intergenerational research and engagement. *Journal of Global Ageing, 1*(2), 219–237.

Thomas, M. J., Sorvala, L., Williams, A., Singleton, A. Maddock, C., Morgan, D., Murray, T., & Musselwhite, C. (in press). Co-creating a climate comic book: Reflections on using comics in intergenerational research and engagement. *Journal of Global Ageing*.

United Nations. (2020). *Analytical study on the promotion and protection of the rights of persons with disabilities in the context of climate change*. Report of the Office of the United Nations High Commissioner for Human Rights. Retrieved October 3, 2024, from https://www.ohchr.org/en/climate-change/impact-climate-change-rights-persons-disabilities

United Nations. (2022). *World population prospects*. United Nations. Retrieved October 3, 2024, from https://www.un.org/development/desa/pd/sites/www.un.org.development.desa.pd/files/wpp2022_summary_of_results.pdf

Van Cauwenberg, J., De Bourdeaudhuij, I., Clarys, P., de Geus, B., & Deforche, B. (2019). E-bikes among older adults: Benefits, disadvantages, usage and crash characteristics. *Transportation*, *46*, 2151–2172.

Walker, I., Tapp, A., & Davis, A. (2023). Motonormativity: How social norms hide a major public health hazard *International Journal of Environment and Health*, *11*(1), 21–33.

Watkins, S., & Musselwhite, C. B. A. (2024). Recognised cognitive biases: How far do they explain transport behaviour? *Journal of Transport & Health*, *40*, 101941.

Willoughby, M., Kipsaina, C., Ferrah, N., Blau, S., Bugeja, L., Ranson, D., & Ibrahim, J. E. (2017). Mortality in nursing homes following emergency evacuation: A systematic review. *Journal of the American Medical Directors Association*, *18*(8), 664–670.

World Health Organization. (2012). *Environmental health inequalities in Europe: Assessment report*. WHO European Centre for Environment and Health. WHO Regional Office for Europe, Copenhagen. Retrieved October 3, 2024, from http://www.euro.who.int/en/publications/abstracts/environmental-health-inequalities-in-europe-assessment-report

World Health Organization. (2020). *Healthy ageing and functional ability*. Retrieved October 3, 2024, from https://www.who.int/news-room/questions-and-answers/item/healthy-ageing-and-functional-ability

World Health Organization. (2021). *Global report on ageism*. World Health Organization. Retrieved October 3, 2024, from https://www.who.int/publications/i/item/9789240016866

CHAPTER 18

PUBLIC PARTICIPATION AS A MEANS OF LEGITIMISING NET ZERO TRANSPORT: CHALLENGES AND OPPORTUNITIES

Claus Hedegaard Sørensen[a], Tom Rye[b,c] and Lisa Hansson[b]

[a]*The Swedish National Road and Transport Research Institute (VTI), Sweden*
[b]*Molde University College – Specialized University in Logistics, Norway*
[c]*Urban Planning Institute of Slovenia, Slovenia*

ABSTRACT

Increasing the political legitimacy of policies to bring about net zero transport is a core challenge for the transformation of the transport sector. This is the case not least because the policies required will probably imply less and slower mobility, challenging the high-mobility society to which many people have become accustomed. One way to increase the legitimacy of these measures could be by public participation which, however, is often subject to a number of challenges, four of which are discussed in this chapter, namely: Who is the public(s)? How to gain awareness? Does the public(s) possess relevant knowledge? Should consensus be an ideal? We conclude by suggesting representative and deliberative public participation as a way forward that could contribute to solving these and other challenges.

Keywords: Public participation; policy legitimacy; the public; awareness; knowledge; consensus

Towards Transport Net Zero
Transport and Sustainability, Volume 20, 269–281

Published under exclusive licence by Emerald Publishing Limited
ISSN: 2044-9941/doi:10.1108/S2044-994120250000020018

INTRODUCTION

A rapid transformation to global net zero of carbon dioxide emissions from human activity is necessary (Kemp et al., 2022). This also applies to the transport sector, where the Intergovernmental Panel on Climate Change (IPCC) has identified the need for a 'paradigm shift that ensures prioritisation of high-accessibility transport solutions that minimise the amount of mobility required to meet people's needs and favours transit and active transport modes' (IPCC, 2022, p. 1057). For those who suffer from mobility poverty (see, for example, Martens, 2016), a shift of this kind will likely improve mobility options, but for many people, a shift of this character implies less and slower mobility compared to their current travel patterns. Less, because there is a need for people to make shorter and possibly fewer trips, and slower, because of a shift to more sustainable modes which cannot necessarily compete in terms of speed and comfort with aero- and automobility.

In particular in the Global North, but also in affluent segments of society in the Global South, the population has become accustomed to a high-mobility society (Kaufmann & Viry, 2015) and a way of life characterised by less and slower mobility might generally seem unattractive to them (Cass et al., 2023; Cohen & Gössling, 2015), implying a probable lack of support for and legitimacy of policies that would move us towards a low-carbon-mobility society. Lack of legitimacy makes it difficult for governments to propose and implement such policies.

On the other hand, public participation could contribute to increasing the legitimacy of these policies and is the focus of this chapter. It can contribute if it enables the knowledge, norms, and values of the population to be incorporated in the policy proposal developed. Community and local knowledge highlighted by participatory efforts may improve the proposal and lead to support simply because it seems a good and appropriate policy. However, public participation can also contribute to legitimacy by means of the process and methods applied when developing and implementing the policy in question. This could be the case where the participatory process allows various groups in the population to participate, and as a result people feel that decision-makers are listening and understand their concerns, ideas, and suggestions. Even if large parts of the population are not participating themselves, legitimacy can still be enhanced, if people feel that they are well represented by others who do participate (Sørensen et al., 2021, 2023). Participation in decision-making processes by stakeholders (i.e., interest organisations and companies) can also contribute to legitimacy but is not the focus of this chapter.

All in all, public participation may well contribute to legitimation of policies of less and slower mobility that are needed to achieve net zero emissions, and there is also much experience that illustrates the difficulties of gaining legitimacy if the population is not involved in the process (e.g., Campbell, 2023; Tønnesen et al., 2023). However, the literature also suggests many challenges in participatory processes that might limit their potential to enhance legitimacy (Sørensen et al., 2023) and that could even worsen the outcome in terms of net zero (Rye, 2023). Four such challenges will form the main body of this chapter, here formulated as questions: Who are the public(s)? How to gain awareness? Do the public(s) possess relevant knowledge? Should consensus be an ideal?

This chapter is based on empirical examples and discussions in an anthology, *Public Participation in Transport in Times of Change* (Hansson et al., 2023), which we edited, but is also informed by other literature on public participation, mostly from the field of transport and mobility. We acknowledge that the standpoint from which we write reflects primarily experience of public participation in the Global North and particularly in western Europe and North America, and that experience in other countries may be significantly different, particularly with respect to power relations and levels of conflict and their impact on participation activities.

WHO IS THE PUBLIC(S)?

Who are the public(s)? (We would use the plural here, as in other political science literature (e.g., Toepfl & Piwoni, 2018) to emphasise that citizens are not one, homogeneous group.) It is a simple question, though not one that is easy to answer, even if it is highly relevant when the topic under consideration is public participation. The challenge seems particularly great when it comes to climate mitigation which requires coordination on multiple scales as well as multilevel governance (Sprain, 2016).

In the Aarhus Convention dealing with public participation in decision-making on environmental matters, 'the public concerned' is defined as 'the public affected or likely to be affected by, or having an interest in, the environmental decision-making' (Aarhus Convention, 1998, Article 2). Taking that definition at face value as an approach to determine who the publics are, it could include generations not yet born, as well as people in countries far away that are affected by our greenhouse gas emissions. The challenge of integrating these publics in specific climate mitigation efforts is obvious. When dealing specifically with the transport sector, another challenge of defining the public occurs. Policies to advance net zero transport most likely will affect both the publics living in an area and those visiting the area either for leisure or passing through to reach other destinations. Thus, the mobile nature of the field of transport constitutes a particular challenge for public participation that makes it different from many other sectors (Sørensen et al., 2023).

When it comes to concrete participatory processes, answering the question of who the publics are implies other concerns. In many countries, participatory processes are legally required when planning changes in infrastructure, such as road enlargement projects. Socio-economic groups, particularly those with political and financial resources and who are affected by the changes, are often active in these processes. However, it is important to achieve representativeness in participation, for example, with respect to age, gender, and social class and, in the case of transport, also across users of different modes of transport. This is important, since different ages, genders, classes, and transport users are often all concerned members of the public and might contribute different perspectives (Wild et al., 2018).

If challenges of defining the relevant publics and ensuring that they participate in decision-making on net zero transport are not addressed, several problems emerge. First, absent or insufficient participation by the affected publics implies that they are not involved in problem definition and thus not involved in raising their perspectives, understanding, and interests. Second, these publics do not contribute to the actual decisions made, regardless of whether the participation is characterised by consultation or actual decision-making agency (Arnstein, 1969). And finally, the possible policy legitimacy provided by participative processes does not include the excluded publics, and hence, the decision might suffer from lack of legitimacy and support.

Drawing on the literature on environmental justice, ideas of distributional (distribution of costs and benefits), recognition (complete and equal political rights), and procedural (participation and distribution of power) justice are all important in the definition of the publics in public participation (Jenkins et al., 2016). If those organising public participation on questions of policies to achieve net zero transport do not resolve the question of who the concerned publics are in determining the problem, influencing actual decisions and in legitimising them, then attempts to involve the public might rather strengthen existing power relations and exclude certain publics (Glimmerveen et al., 2022; Sprain, 2016). In that case, public participation will neither contribute to distributive nor to procedural justice and runs the risk that net zero policies exacerbate inequity and injustice and thus lack legitimacy. Some of the popular protests against transport policies related to climate change mitigation, such as the *Gilets Jaunes* in France, the Fuel Rebellion in Sweden, and No to Toll Roads in Norway, could be seen as a reaction to policies characterised by perceived inequity and injustice.

Considering who constitutes the relevant publics is, therefore, a core question for both practice and research, and new methods and ways to include excluded publics need much more attention.

HOW TO GAIN AWARENESS?

Creating awareness of participatory processes is a precondition for participation. Awareness also signals that people have a role to take part in democratic processes and societal building in general. A focus group study on public participation by Lowndes et al. (2001) shows that citizens had little awareness of their rights and opportunities to actually influence and participate in various activities related to the jurisdiction of their own municipal government, for example.

There are several challenges at the (local) government level when it comes to raising awareness of public participation processes. One central factor is how to identify relevant information channels that will reach a broad and diverse group of people in the affected area. Municipalities' increased use of information and communication technology (ICT) services has enhanced their visibility online and may contribute to raising awareness. However, several challenges related to ICT have also been identified in previous research. Some municipalities lack the resources to provide relevant and transparent information on their websites

(Simelio et al., 2019). The quality of the websites can be poor, and these can be difficult to navigate and/or the information is presented poorly (Lee-Geiller & Lee, 2019). There may also be language barriers that lead to exclusion of certain groups (Harvey & Brazier, 2022), and if few people regularly visit government webpages, it is difficult to reach a broader range of citizens.

A complement to the municipal page is using social media channels (Simelio et al., 2019). However, these two methods combined still do not solve the challenge of digital illiteracy, where people cannot use, or do not have access to, internet services. Budding et al. (2018) point out that it is primarily the working population that benefits from e-government adaptation. This group uses municipal e-channels to gain information about activities or services, instead of physically visiting municipal offices.

There are also challenges in informing and getting people to understand the relevance of participating in various processes. The literature shows examples of marginalised groups that do not feel welcome to participate in these kinds of processes. Furthermore, there are also groups that have a low level of trust in public institutions (Laurian, 2009; Lowndes et al., 2001). In addition, there are challenges related to lack of time and other resources. To overcome such obstacles, various municipal outreach activities have been suggested in the literature, for example, conducting workshops or meetings in the context where excluded groups normally spend time (OECD, 2022). Events in public spaces can also be a way to reach various groups, such as a bus tour by politicians and planners, where stops are used to inform users about plans for and developments of the system and to get user feedback (OECD, 2022).

Mass media also plays a critical role as a mediator of communication between politicians, planners, and the public (Innes, 1998). Media coverage can provide information on ongoing planning activities as well as increase interest in planning issues (Innes & Booher, 2003; Iyengar & Kinder, 2010). Mass media can also summarise statistics and political initiatives in a more comprehensible and user-friendly manner than official public documents. However, there can be challenges when mass media is seen as a communication channel for raising awareness about public participation. Mass media has an agenda-setting role, which can influence decisions on what to publish and lead to the prioritisation of certain issues over others. Media can also influence the public through choice of language, tone, and framing (Ninan et al., 2022). For example, a newspaper may provide positive and informative reporting on the climate change benefits of electric vehicles, or it may focus on challenges such as a range of anxiety or safety issues (Brüggemann, 2017).

Media can further use its role by influencing who is represented in the media and who is ignored (Ninan et al., 2022). Media coverage of new bicycle infrastructure projects may focus on the voices of enthusiastic cyclists or urban planners but leave out the opinions of residents who may be concerned about changes in the area, or those who do not use bikes as their primary mode of transportation (Wild et al., 2018). If the media choose not to include the perspectives of a varied range of people, this can result in a biased narrative that does not accurately represent diverse viewpoints and needs (Ninan et al., 2022). On the other hand, by covering public forums, community events, and citizen-led initiatives, newspapers can contribute to creating a narrative of active public participation. This gives visibility to initiatives and community members who are actively involved in discussions and sustainable transportation projects and may also inspire other community members to get involved in sustainable transportation initiatives (Nello-Deakin, 2023).

Raising awareness of public participation may also come from citizen-led, bottom-up initiatives. Drawing on the literature inspired by Lefebvre (1967) and 'Right to the City', there are several examples where individuals and groups take an activist role to influence local government processes (Harvey, 2015). A typical example is guerilla-painted bike lanes to indicate space on the road for cyclists. Other examples are flash-mobs on streets or bicycle protests on car lanes. From a municipal planning perspective, such initiatives may lead to conflicts between citizen groups and municipal planners and delay processes (Stevens & Dovey, 2022). However, whether activism per se is a challenge is largely a matter of perspective.

DOES THE PUBLIC(S) LACK RELEVANT KNOWLEDGE?

Public participation processes depend on those who take part having knowledge of what is proposed, the reasons for it, and its likely impacts. Authors such as Vigar (2017) point out that in planning processes there are various types of knowledge, such as embodied/local knowledge, technical knowledge, practice-centred knowledge about what actually works, and political knowledge. It is important to recognise that no one group in the participation process has a monopoly on all these different forms of knowledge.

In addition, there is clear evidence that greater knowledge of the causes of and possible responses to climate change leads to greater support for those responses (Pickering et al., 2022). With regard to transport and mobility initiatives, Gaunt et al. (2006) show how greater knowledge of a proposed congestion charging scheme in Edinburgh, UK, was correlated with greater likelihood to vote in favour of that scheme in a referendum that was held on the proposal. Maltese et al. (2023) also show that a significant barrier to the involvement of a wider public in participation about local and regional freight transport policy in the context of sustainable urban logistics plans (SULPs) is lack of knowledge. Syklistenes Landsforening (2024) shows how a lack of knowledge about the nature and objectives of new cycleway schemes in Oslo generated considerable protest against them, thus making the public participation process confrontational and difficult to optimise.

Returning to Vigar's (2017) work regarding different types and levels of knowledge of different actors in participation, it is important to note that knowledge is not something that is necessarily brought to the participation process by those who organise that process and who are the proponents of the scheme. It is perfectly possible that people or stakeholders who have been invited to participate will also bring knowledge about the scheme that can be used to improve it or who will in some cases have better knowledge of design standards or funding possibilities than those who are designing the scheme. Local people may also have a deeper understanding of the history of a location than those who are designing the scheme that will be located there. This two-way nature of knowledge is very important to appreciate (Martin, 2008).

There is also a recognised phenomenon that people will tend to believe the information and opinions about a new transport initiative that they hear first, regardless of its accuracy. As Timmons et al. (2023, p. 28) state:

> [...] all of this takes place in a context where people are most strongly influenced by the information and arguments that they hear first and who they hear the arguments from. Once opinions have formed, even when based on poor information or superficial arguments, they can be resistant to new information that challenges them.

Thus, it is not only whether knowledge is actually correct or true but also when and by whom it is delivered, that influences the effect of that knowledge on the participation process. Levels of trust in the message-giver will critically affect whether knowledge is accepted and, hence, whether the need for the (transport) policy or measure in question is also accepted.

A further aspect related to knowledge and the message-giver is that knowledge about the policy or measure proposed is often couched in very technical terms by the proponents of the policy or measure; such knowledge may be difficult for those taking part in participation processes to understand and/or relate to, and their understanding of the policy or measure may be radically different. Horlick-Jones et al. (2007) argue that public participation processes need to recognise these different constructions of knowledge of and about the proposed policy or measure, while at a more basic level, it is obvious that proposals need to be communicated in an easily comprehensible way if they are not to be misunderstood. However, the multiple ways in which people can now gain understanding (or misunderstanding) of a proposed policy or measure (as discussed in the previous section, particularly with respect to media and social media) make it doubly challenging to communicate knowledge of the scheme in a clear way.

Overall, then it can be seen that there are many ways in which knowledge, or a lack of it, can fundamentally influence the level and nature of participation in participatory processes and also the outcome of these processes. The way that this knowledge is constructed, shared, and understood is critical to the effectiveness of the participation process and therefore potentially to the outcome of the proposed policy or measure.

SHOULD CONSENSUS BE AN IDEAL?

In the literature related to public consultation and participation in transport, particularly in more normative texts (e.g., Lindenau & Boehler-Baedeker, 2014; Rupprecht Consult, 2019), there is a strong emphasis on these processes achieving a consensus about the transport measures to be implemented. Through a process of deliberation, it is argued, a consensus about the optimal set of measures will be reached. However, for transport policies to be transformative, measures are required that will result in both winners and losers – a good example is the reduction in space for parked or moving private cars in order to create space for other modes. Clearly, this measure results in short-term negative impacts for some car users, and not all car drivers are therefore likely to support it when participating in public engagement processes. Rye (2023) also cites empirical evidence of transport planners in the United Kingdom who in practice feel under political pressure to employ public participation as a means to achieve consensus on planned measures even as they believe that such consensus is not possible. On the other hand, Hrelja et al. (2024) cite evidence of politicians in other contexts who, pointing to their electoral mandate and the manifesto behind this, appear willing to tolerate a situation where there is some lack of consensus and some active opposition to transformative transport projects. The key argument here, to which we return in the next section, concerns how far political legitimacy is conferred by achieving consensus in participation processes, compared to other means of achieving political legitimacy.

In case studies of public participation, various authors have shown that, in these cases at least, consensus is not a precondition of delivering transformative change in transport. Song (2023) studies the use of citizen ballots in California as a way to achieve support for and ultimately deliver transport policies that are transformative for people who are typically underrepresented and lack influence in conventional public participation processes. Clearly, in a ballot, there are some participants who agree with the proposition and some who do not, so the process by definition is not consensual. Barata et al. (2023) show how citizen participation is transformed into citizen action in the absence of inclusive channels for public participation in a favela in Rio de Janeiro. Finally, Rye (2023) considers whether more public participation in local and regional transport planning leads to 'better' outcomes. In the cases reviewed, he finds that more consensual public participation processes are not necessarily associated with 'better' outcomes in terms of achieving the objectives of the transport scheme that is finally implemented and how long that scheme takes to implement. Based on the work by Priestley (2005), Rye compares the participation processes for planning new tram lines in Lyon, France and Edinburgh, UK. The latter was one in which there was a wide range of participation possibilities over a long period of time and where there was a political imperative to try to use the process to achieve a consensus about the scheme, from its objectives, right through to detailed design. Such an emphasis on consensus is not explicitly part of public participation for new tram lines in Lyon, where many issues that would be the subject of public participation in the UK city are decisions made directly by technical staff. The lines in Lyon are implemented more rapidly than in Edinburgh, but the participation process in Edinburgh has little impact on the original plans for the scheme, and public satisfaction with the local transport system is higher in Lyon than in Edinburgh, which raises questions about the value of attempting to achieve consensus.

DISCUSSION AND CONCLUSIONS

The point of departure for this chapter was that public participation under certain circumstances might contribute to providing legitimacy for policies aimed at delivering less and slower mobility to help bring about net zero transport. It is important to consider how to introduce and strengthen public participation, and we have discussed four topics that can challenge public participation and thus its ability to provide legitimacy. We now will discuss some possible solutions to the challenges but first some words on legitimacy.

Literature suggests that the legitimacy of policies depends on several factors. Two of these, on which we focus here, are normative legitimacy and procedural legitimacy.[1] Normative legitimacy describes how well the policy in question is anchored in the political values and norms of the population (Smoke, 1994), while procedural legitimacy has to do with the way that governing bodies choose to develop and implement the policy (Wallner, 2008). For example, taking into account different perspectives from various stakeholders and the population more broadly might help to better align a proposed policy with the values and norms of the population and thus strengthen its normative legitimacy. And if the processes underlying the development of the proposal are characterized by negotiation, dialogue, and participation, and its implementation is finally decided with a large majority in favour in a democratically elected assembly, this may all contribute to procedural legitimacy. Obviously, there can be a trade-off between the two types of legitimacy. For example, people who disagree with the actual content of a policy (thus for them it has limited normative legitimacy), might nonetheless find it legitimate if it is developed and implemented using an open and inclusive process (thus giving it high procedural legitimacy) – or vice versa (Kronsell, 2013).

To recap, our four challenges were: who the public(s) are, how to raise awareness of participation, whether the public(s) possess relevant knowledge, and whether consensus should be an ideal outcome of participation. There are no easy solutions to these challenges, but a promising way forward for public participation contributing to legitimacy could be what has been described as representative and deliberative participation (OECD, 2020). The idea is that a group of people are invited to a participation process, ensuring that they are representative of the overall population in terms of characteristics such as age, housing tenure, residential location, profession, gender, class, car ownership, and so on. Furthermore, a deliberative approach is adopted, where the group has considerable time to meet, learn, listen, and discuss. This approach has been applied not least in the form of climate citizen assemblies at national and local levels, for example, in France, Denmark, and Scotland.

With respect to the challenges discussed in this chapter, this method has some advantages compared to other forms of public participation. First, as regards the issue of who the public(s) are, the focus on ensuring representativeness of the people involved ensures that one group does not dominate – it will include, for example, residents as well as visitors to the area in question, both weaker and stronger socio-economic groups and people with and without a car. Second, regarding the problem that people have limited time and resources to take part in public participation, an advantage of the representative and deliberative approach is that the entire affected population is not expected to participate but only a small group representing the wider population. So, for example, in the citizen climate assembly in France, there were 150 citizens (Giraudet et al., 2022) and in Denmark, 99 (Sørensen & Isaksson, 2021). Thus, while it may be time-consuming to participate, only a small group is expected to do so.

Third, when it comes to the participants' knowledge of the – often rather technical – field of transport, the deliberative aspect of the approach means that the participants spend some time acquiring that knowledge by means of invited presentations, which they themselves can choose to request; some degree of 'education' takes place in the process. Fourth and finally, while a deliberative approach usually aims for consensus as an ideal, in practice, decisions can be taken by (qualified) majority voting for each recommendation, and/or minority statements can also be an outcome.

An example of representative and deliberative citizen participation within the field of transport is from the Danish capital, Copenhagen, where the use of public space and traffic conditions in the medieval area of the city were the topics of a citizen assembly. It consisted of 36 citizens, differing in age, gender, profession, and car ownership, of whom 75% were residents of the medieval area and 25% from other parts of the City of Copenhagen. The assembly met six times, beginning with many presentations about the issues involved but later focusing more on discussions and finally making recommendations to the City Council's politicians. Consensus was not a requirement and the process allowed for minority statements, but in the end, some degree of consensus was reached by incorporating different viewpoints in the discussions, and the option of minority statements was not used. Among more recommendations, the assembly suggested to reduce the number of on-street parking spaces in the medieval area (from 1,050 to 300). The assembly to some extent contributed to normative legitimacy by proposing recommendations that were well in accordance with concerns and wishes of the citizens living in or outside the area. Further, the diversity and representativeness of the people that participated in the assembly provided for procedural legitimacy vis-à-vis the politicians in the City Council and perhaps also to some extent in the broader population, though limited media coverage probably reduced the latter effect.

Public participation is no Columbus' Egg for giving legitimacy to policies for less and slower mobility and net zero transport. Similarly, representative and deliberative participation does not deal with all the challenges of public participation we highlight in this chapter. Even if representative and deliberative participation is used, there will be groups excluded, awareness and time will be a challenge, along with lack of knowledge and – in some cases – the pursuit of an ideal of consensus which in fact often is impossible to achieve. The distribution of power and power relations with public participation processes are of course hugely important, but we have not touched on them here directly because the topic would merit at least another book chapter on its own. Nonetheless, it is hard to imagine a future of less and slower mobility and net zero transport, unless policies favouring such a development are considered legitimate in the eyes of the population. The legitimacy of various net zero policies and the mechanisms to gain legitimacy have to be considered carefully by policymakers at all levels. Involving the public is among these mechanisms, and in particular, we suggest representative and deliberative approaches to public participation as a promising method that should be the object of more experimentation and research.

NOTE

1. Other, partly overlapping factors are mentioned in the literature, e.g., substantive legitimacy, cognitive legitimacy, output legitimacy, input legitimacy, and throughput legitimacy (Kronsell, 2013; Smoke, 1994; Wallner, 2008).

REFERENCES

Aarhus Convention. (1998). *Convention on access to information, public participation and decision-making and access to justice in environmental matters.* https://unece.org/environment-policy/public-participation/aarhus-convention/text

Arnstein, S. R. (1969). A ladder of citizen participation. *Journal of the American Institute of Planners, 35*(4), 216–224.

Barata, A. F., Jones, T., & Brownill, S. (2023). Understanding the multiple roles of participation in urban mobility: An investigation of spaces for participation in Rio de Janeiro, Brazil. In L. Hansson, C. H. Sørensen, & T. Rye (Eds.), *Public participation in transport in times of change* (Vol. 18, pp. 33–51). Emerald Publishing Limited.

Brüggemann, M., & Engesser, S. (2017). Beyond false balance: How interpretive journalism shapes media coverage of climate change. *Global Environmental Change*, *42*, 58–67.

Budding, T., Faber, A. S. C., & Gradus, R. H. J. M. (2018). Assessing electronic service delivery in municipalities: Determinants and financial consequences of e-government implementation. *Local Government Studies*, *44*(5), 697–718.

Campbell, M. (2023). Messaging, emergencies, and public discontent: Implementing active travel initiatives during COVID-19. In L. Hansson, C. H. Sørensen, & T. Rye (Eds.), *Public participation in transport in times of change* (Vol. 18, pp. 117–136). Emerald Publishing Limited.

Cass, N., Büchs, M., & Lucas, K. (2023). How are high-carbon lifestyles justified? Exploring the discursive strategies of excess energy consumers in the United Kingdom. *Energy Research & Social Science*, *97*, 102951.

Cohen, S. A., & Gössling, S. (2015). A darker side of hypermobility. *Environment and Planning A: Economy and Space*, *47*(8), 166–1679.

Gaunt, M., Rye, T., & Ison, S. (2006). Gaining public support for congestion charging: Lessons from referendum in Edinburgh, Scotland. *Transportation Research Record*, *1960*(1), 87–93.

Giraudet, L. G., Apouey, B., Arab, H., Baeckelandt, S., Begout, P., Berghmans, N., Blanc, N., Boulin, J.-Y., Buge, E., Courant, D., Dahan, A., Fabre, A., Fourniau, J.-M., Gaborit, M., Granchamp, L., Guillemot, H., Jeanpierre, L., Landemore, H., Laslier, J.-F., …, & Tournus, S. (2022). 'Co-construction' in deliberative democracy: Lessons from the French Citizens' Convention for Climate. *Humanities and Social Sciences Communications*, *9*(1), 1–16.

Glimmerveen, L., Ybema, S., & Nies, H. (2022). Who participates in public participation? The exclusionary effects of inclusionary efforts. *Administration & Society*, *54*(4), 543–574.

Hansson, L., Sørensen, C. H., & Rye, T. (2023). What is public participation in transport in times of change? In L. Hansson, C. H. Sørensen, & T. Rye (Eds.), *Public participation in transport in times of change* (Vol. 18, pp. 3–12). Emerald Publishing Limited.

Harvey, D. (2015). The right to the city. In R. T. LeGates & F. Stout (Eds.), *The city reader* (pp. 314–322). Routledge.

Harvey, M., & Brazier, D. (2022). E-government information search by English-as-a Second Language speakers: The effects of language proficiency and document reading level. *Information Processing & Management*, *59*(4), 102985.

Horlick-Jones, T., Rowe, G., & Walls, J. (2007). Citizen engagement processes as information systems: The role of knowledge and the concept of translation quality. *Public Understanding of Science*, *16*(3), 259–278.

Hrelja, R., Rye, T., & Hansson, L. (2024, April). Understanding politicians' positions about contentious car restrictive measures: Public protest and ways to build support for them. Paper to MobilTUM conference, Munich.

Innes, J. E. (1998). Information in communicative planning. *Journal of the American Planning Association*, *64*(1), 52–63.
Innes, J. E., & Booher, D. E. (2003). Collaborative policymaking: Governance through dialogue. In M. A. Hajer & H. Wagenaar (Eds.), *Deliberative policy analysis: Understanding governance in the network society* (pp. 33–59). Cambridge University Press.
IPCC. (2022). *Climate change 2022. Mitigation of climate change*. Retrieved January 22, 2024, from https://www.ipcc.ch/report/ar6/wg3/downloads/report/IPCC_AR6_WGIII_FullReport.pdf
Iyengar, S., & Kinder, D. R. (2010). *News that matters: Television and American opinion*. University of Chicago Press.
Jenkins, K., McCauley, D., Heffron, R., Stephan, H., & Rehner, R. (2016). Energy justice: A conceptual review. *Energy Research & Social Science*, *11*, 174–182.
Kaufmann, V., & Viry, G. (2015). High mobility as social phenomenon. In G. Viry and V. Kaufmann (Eds.), *High mobility in Europe: Work and personal life* (pp. 1–15). Palgrave Macmillan UK.
Kemp, L., Xu, C., Depledge, J., Ebi, K. L., Gibbins, G., Kohler, T. A., Rockström, J., Scheffer, M., Schellnhuber, H. J., Steffen, W., & Lenton, T. M. (2022). Climate endgame: Exploring catastrophic climate change scenarios. *Proceedings of the National Academy of Sciences*, *119*(34), e2108146119.
Kronsell, A. (2013). Legitimacy for climate policies: Politics and participation in the Green City of Freiburg. *Local Environment*, *18*(8), 965–982.
Laurian, L. (2009). Trust in planning: Theoretical and practical considerations for participatory and deliberative planning. *Planning Theory & Practice*, *10*(3), 369–391.
Lee-Geiller, S., & Lee, T. D. (2019). Using government websites to enhance democratic E-governance: A conceptual model for evaluation. *Government Information Quarterly*, *36*(2), 208–225.
Lefebvre, H. (1967). Le droit à la ville. *L'Homme et la Société*, *6*(1), 29–35.
Lindenau, M., & Boehler-Baedeker, S. (2014). Citizen and stakeholder engagement: A precondition for sustainable urban mobility. *Transportation Research Procedia*, *4*, 347–360.
Lowndes, V., Pratchett, L., & Stoker, G. (2001). Trends in public participation: Part 2 – Citizens' perspectives. *Public Administration*, *79*(2), 445–455.
Maltese, I., Marcucci, E., Gatta, V., Sciullo, A., & Rye, T. (2023). Challenges for public participation in sustainable urban logistics planning: The experience of Rome. In L. Hansson, C. H. Sørensen, & T. Rye (Eds.), *Public participation in transport in times of change* (pp. 77–95). Emerald Publishing Limited.
Martens, K. (2016). *Transport justice: Designing fair transportation systems*. Routledge.
Martin, G. P. (2008). 'Ordinary people only': Knowledge, representativeness, and the publics of public participation in healthcare. *Sociology of Health & Illness*, *30*(1), 35–54.
Nello-Deakin, S. (2023). 'They are deconstructing the city': A typology of media narratives on tactical urbanism. *Journal of Urbanism: International Research on Placemaking and Urban Sustainability*, 16, 1–20.
Ninan, J., Mahalingam, A., & Clegg, S. (2022). Power in news media: Framing strategies and effects in infrastructure projects. *International Journal of Project Management*, *40*(1), 28–39.
OECD. (2020). *Innovative citizen participation and new democratic institutions: Catching the deliberative wave*. Organisation for Economic Co-operation and Development. https://www.oecd-ilibrary.org/governance/innovative-citizen-participation-and-new-democratic-institutions_339306da-en
OECD. (2022). *OECD guidelines for citizen participation processes*. OECD Publications Centre. https://doi.org/10.1787/f765caf6-en
Pickering, J., Hickmann, T., Bäckstrand, K., Kalfagianni, A., Bloomfield, M., Mert, A., Ransan-Cooper, H., & Lo, A. Y. (2022). Democratising sustainability transformations: Assessing the transformative potential of democratic practices in environmental governance. *Earth System Governance*, *11*, 100131.
Priestley, A. (2005). *How to fast-track Edinburgh's trams. A study of transport project delivery processes in Lyon and Edinburgh with recommendations for improvement of the Scottish system* [Unpublished MSc dissertation]. Edinburgh Napier University.
Rupprecht Consult. (2019). *Guidelines for developing and implementing a sustainable urban mobility plan*. Rupprecht Consult Forschung & Beratung GmbH.

Rye, T. (2023). Does more public participation in local and regional transport planning lead to 'better' outcomes? In L. Hansson, C. H. Sørensen, & T. Rye (Eds.), *Public participation in transport in times of change* (pp. 157–178). Emerald Publishing Limited.

Simelio, N., Ginesta, X., de San Eugenio Vela, J., & Corcoy, M. (2019). Journalism, transparency and citizen participation: A methodological tool to evaluate information published on municipal websites. *Information, Communication & Society*, *22*(3), 369–385.

Smoke, R. (1994). *On the importance of policy legitimacy* (pp. 97–110). Political Psychology.

Song, L. (2023). Mobilising for transit-oriented communities in Los Angeles. In L. Hansson, C. H. Sørensen, & T. Rye (Eds.), *Public participation in transport in times of change* (pp. 15–31). Emerald Publishing Limited.

Sprain, L. (2016). Paradoxes of public participation in climate change governance. *The Good Society*, *25*(1), 62–80.

Stevens, Q., & Dovey, K. (2022). *Temporary and tactical urbanism: (Re)assembling urban space*. Taylor & Francis.

Syklistenes Landsforening. (2024). *Fra omstridte tiltak til bred støtte* (*From controversial measures to broad support*). Unpublished handbook on public consultation and implementation of contentious transport measures. Norwegian Cyclists Association.

Sørensen, C. H., & Isaksson, K. (2021). *Omställning till hållbar mobilitet: Vilka roller kan smart mobilitet, medborgardeltagande och coronakrisen spela?* Statens väg-och transportforskningsinstitut.

Sørensen, C. H., Hansson, L., & Rye, T. (2021). Medborgarnas roll i transportpolitiken och i omställningen mot ett hållbart transportsystem. In L. Eriksson, K. Isaksson, & J. Witzell (Eds.), *På väg mot hållbar omställning? Kunskap, makt och mening i nationell transportplanering*. Linnefors Förlag.

Sørensen, C. H., Hansson, L., & Rye, T. (2023). The transformational potential of public participation in transport. In L. Hansson, C. H. Sørensen, & T. Rye (Eds.), *Public participation in transport in times of change* (Vol. 18, pp. 181–194). Emerald Publishing Limited.

Timmons, S., Andersson, Y., McGowan, F., & Lunn, P. (2023). *Using behavioural science to design and implement active travel infrastructure: A narrative review of evidence*. SSRN 4365989.

Toepfl, F., & Piwoni, E. (2018). Targeting dominant publics: How counterpublic commenters align their efforts with mainstream news. *New Media & Society*, *20*(5), 2011–2027.

Tønnesen, A., Krogstad, J. R., & Christiansen, P. (2023). Governing urban transport packages in Norway: Understanding conditions for public participation. In L. Hansson, C. H. Sørensen, & T. Rye (Eds.), *Public participation in transport in times of change* (pp. 137–156). Emerald Publishing Limited.

Vigar, G. (2017). The four knowledges of transport planning: Enacting a more communicative, trans-disciplinary policy and decision-making. *Transport Policy*, *58*, 39–45.

Wallner, J. (2008). Legitimacy and public policy: Seeing beyond effectiveness, efficiency, and performance. *Policy Studies Journal*, *36*(3), 421–443.

Wild, K., Woodward, A., Field, A., & Macmillan, A. (2018). Beyond 'bikelash': Engaging with community opposition to cycle lanes. *Mobilities*, *13*(4), 505–519.

CHAPTER 19

COVID-19, PANDEMICS, AND DISRUPTION

Corinne Mulley[a] and Maria Attard[b]

[a] *University of Sydney, Australia*
[b] *University of Malta, Malta*

ABSTRACT

The 2020 COVID-19 pandemic had a significant impact on transport systems and mobility. This large-scale disruption led to several initiatives that changed temporarily, and even permanently in some cases, cities and people's behaviour. This chapter looks at this disruptive event and discusses the various initiatives that occurred in different sectors of transport operations, highlighting their potential contribution to climate change mitigation and adaptation efforts. Some of the more positive changes have the potential to assist cities in achieving climate goals. There is a clear urgency for such actions, and by looking at the lessons learnt from the COVID-19 pandemic, we can aid in achieving a more effective pathway to net zero.

Keywords: Pandemic; COVID-19; climate change; travel behaviour; policy pathways; disruption

INTRODUCTION

This chapter explores the theme of disruption in transport by looking at the effects of the recent COVID-19 pandemic (2020) on mobility, and the potential impacts of sustainable transport actions on climate change mitigation and adaptation. It begins by defining disruption and reflecting on the interrelated issues of transport as

Towards Transport Net Zero
Transport and Sustainability, Volume 20, 283–296

Published under exclusive licence by Emerald Publishing Limited
ISSN: 2044-9941/doi:10.1108/S2044-994120250000020019

a vector of contagion and the transport disruptions that have emerged, and are still emerging, due to climate change. COVID-19, as the most recent pandemic, was not only exacerbated by transport issues but also caused many new transport challenges and opportunities. The discussion then sets out a timeline of COVID-19 events from December 2019 through to the recovery of transport systems as the world relaxed following the implementation of vaccines that ameliorated the impact of the pandemic. This includes identifying the changes that cities experienced because of that disruptive event and shows how opportunities transport decision-makers took then could be more widely engaged with now to promote the kind of travel behaviour changes needed to help mitigate and adapt to climate change.

DISRUPTION, CONTAGION, AND DECARBONISATION

Disruption is defined by the Oxford Dictionary as 'the act of stopping something from continuing in the normal way'. This suggests someone or something that changes the course of things or the processes in a system. Transport, like many systems, is made up of various components. The way these components interact impacts on the overall performance of the (transport) system. A change in one of those components has the potential of being disruptive. We return to explore this concept later in this chapter.

Transport as a Vector of Contagion

A pandemic is distinguished from an epidemic and an outbreak by the scale of its impact. An outbreak is characterised by a sudden increase in the number of cases of a disease which may last for a short period (weeks) or longer (up to several years), not necessarily restricted in spatial area to one region or country. In this context, an epidemic is when an infectious disease spreads quickly to a large population. By contrast, a pandemic is a global disease event affecting more people in a larger spatial area, often worldwide, causing more fatalities than an epidemic with associated social and economic disruption (Association for Professionals in Infection Control and Epidemiology, 2014).

There have been several recorded pandemics, each having a severe and deleterious impact on human populations. Typically, the human population has no immunity to the disease, which underpins the rapid spread from one human to another. Examples include the Spanish Flu (1918, causing between 20 and 30 million fatalities), Asian Flu (1957–1958, 1–4 million fatalities), Hong Kong Flu (1968, 1–4 million fatalities), and, more recently, Swine Flu (2009, estimated 294,000 fatalities), the first for which a vaccine was developed. Of course, the most recent was the COVID-19 pandemic (March 2020–May 2023, over 7 million fatalities), and as with Swine Flu, COVID-19 vaccines have been developed and are widely identified as reducing the severity and number of deaths from the disease.

In the case of the COVID-19 pandemic, transport was a critical vector for its diffusion in a short period of time. At first was the transmission of the virus through air transport and subsequently through all modes starting from public transport across and in cities. The reaction to the pandemic was fast, and transport was among the first sectors to be affected. In many cities and states, airports were closed, public transport services decimated, and new hygiene protocols were implemented because of fear of contagion; equally, new individual transport services emerged and urban infrastructure was re-designed in some areas for walking, cycling, and social distancing. With travel being physically restricted in some areas, roads built for cars were left empty and in some cases were reclaimed by wildlife. Tourism travel slowed down almost to a stop, cruise ships struggled with outbreaks of COVID-19, and international travel was dictated by red/green classification lists which highlighted the risks of exposure to the virus for visitors (with some countries banning travel to red and dark red zones). The system that supported freight transport was also affected because of changes in consumption and the rise of home-based internet shopping. This proved challenging for freight and logistics operators that struggled to support the surge in deliveries required.

Transport Disruptions Due to Climate Change

Turning to a rather different source of transport disruption, the *Stern Review on the Economics of Climate Change* (Stern, 2007) reviewed extensive evidence on the impacts of climate change and concluded that 'the scientific evidence is now overwhelming: climate change is a serious global threat, and it demands an urgent global response' (p. viii). The report identifies early action as strongly outweighing the economic costs of not acting. More importantly, it identified climate change as a global issue and that, although the costs are not evenly distributed, actions were needed that did not limit aspirations of growth for either richer or poorer countries. The report summarised four key areas for emissions reduction: emissions trading, technology cooperation, deforestation reduction, and adaptation. The Intergovernmental Panel on Climate Change's (IPCC's) *Sixth Assessment Report* (AR6) went on to attribute the widespread and rapid changes in the atmosphere, ocean, cryosphere, and biosphere to human influence, describing the scale of recent changes across the climate system as unprecedented over many centuries to many thousands of years. The AR6 provided the evidence of observed changes in weather and climate extremes in every region of the globe, with heatwaves, heavy precipitation, droughts, and tropical cyclones all attributed to human influence (IPCC, 2021).

Importantly in the context of this chapter, any decarbonisation effort has to include the transport sector, which contributes significantly to emissions worldwide. Broadly speaking, transport accounts for nearly a quarter of global CO_2 emissions, with nearly half of this coming from passenger travel (Ritchie, 2020; see Chapter 2). Decarbonising transport will need adaptation both to infrastructure and to its use, if the necessary transformation is to be made, especially as the consequences of not doing it early and strongly enough, as suggested by the *Stern Review*, means that what is necessary now is more costly and more painful to do in the future. It is often suggested that transport as a sector is more difficult than others to decarbonise because of the long-term planning associated with major infrastructure developments and the way in which there are complex interactions between significant choices made by citizens – for example, where to live and work – and their transport implications. It is also clear now that technological development, for example, advancements in electrification and information technology (IT), will not on their own be sufficient to deliver necessary reductions in carbon emissions (Creutzig et al., 2018). There will also need to be changes in the way the transport system is used, or in other words, there needs to be behavioural change.

There are some, but not many, examples of disruption to the transport system which engender adaptation or behaviour change. In their study, Marsden et al. (2020) investigate potential disruptions (the impact of winter weather, flooding, and the closure of a major infrastructure link) and the way in which citizens may react to these events, using stated preference studies. Their conclusions do not support the hypothesis that stability in the transport sector makes it difficult to decarbonise. They found that respondents were able to respond to disruption through adaptation using one of seven pathways (remoding, rerouting, retiming, rescheduling, relocating, reallocating, or reducing), although of course the impact of the different adaptation pathways chosen by individuals would have different impacts on the carbon intensity of the replaced trip. Wang, Qu, Yang, Nichol, et al. (2020) look more specifically at climate change risks and implications for planning in the transport sector, and there is specific research in, for example, maritime (Monios & Wilmsmeier, 2020) and rail (Bešinović, 2020) transport (see Chapters 9 and 11). Research into the vulnerability of transport infrastructure is also relatively scant, giving rise to significant gaps in our understanding of how transport systems will perform during severe weather events and disruptions caused by climate change.

A few events, like the Eyjafjallajökull eruption in Iceland in 2010, caused severe disruption, in this case to air travel across a substantial part of Europe (Budd et al., 2011). But until the outbreak of the COVID-19 pandemic, much of the evidence from disruptions was either hypothetical (as with Marsden et al., 2020) or relatively focused on specific events (van Exel & Rietveld, 2009) or local contexts (Dawson et al., 2016). The reach of the pandemic and the way in which transport acts as a vector of disease meant that transport disruption was global and significant. Its impact on behaviour change provides the underpinning of this chapter, which in turn can point to the potential of behaviour change to contribute to the decarbonisation agenda.

TIMELINES OF THE COVID-19 PANDEMIC EVENT – FAST BEHAVIOUR CHANGE, SLOW(ER) RECOVERY

COVID-19 is a coronavirus-based infectious disease caused by the SARS CoV-2 virus. Although coronavirus is a family of infectious respiratory diseases responsible for, among other illnesses, the common cold, COVID-19 brought about the most recent pandemic event because of its rapid spread between humans, worryingly high mortality rates and the non-existence of a vaccine to protect populations. COVID-19 was first detected in Wuhan, China, on 19 December

2019, following the appearance of several pneumonia cases with unknown cause and human-to-human transmission confirmed a month later. By January 2020, cases of the new virus had been identified in Japan, Thailand, and the United States, and the next month saw cases recorded in Europe – in Italy, France, and the United Kingdom – and in Egypt. In early March, a rapid increase in infections that overwhelmed its health system sent Italy into lockdown, a policy that required people to 'stay where they are'. A formal declaration of pandemic was made by the World Health Organization (WHO) on 11 March 2020.

Following this, many countries enforced lockdowns of various degrees of severity on their citizens. Airports closed their operations, with many island states being completely isolated except for emergency (medical) travel. Between June 2020 and the end of the year, various countries began to open their borders again, to ameliorate the economic impacts that were beginning to affect states, communities, and individuals. Upsurges in new infections caused localised lockdowns in this period as 'waves' of cases shifted across space. While Europe and North America were among the first to open, states like Australia and Japan remained closed for much longer, exacerbating the impacts on their transport systems and on movement – and as a result their wider economies and societies – in general.

The first successful vaccines began to appear from late December 2020, from which time the world began to move back towards normality. By March 2022, the rhetoric had changed more to 'living with COVID-19' with most restrictions eased but with outbreaks happening sporadically around the world and continuing to this day, as the virus mutated. As citizens started travelling again, much of the transport industry struggled with staff shortages that plagued operators across modes and operations nationally and internationally. Companies that had made staff redundant due to the removal, or a decline in the number, of services encountered difficulties in re-recruiting at the speed of the recovery. This was also felt by tourism operators in many popular tourist destinations (see, for example, Rejan & Ahn, 2024; Stojčić & Vizek, 2024).

The impact of the pandemic on human life was immense. Table 19.1 provides details of the number of cases and deaths caused by COVID-19 as of March 2022.

Table 19.1. Number of Cases and Deaths Caused By COVID-19 by Region.

Region	Cases	Deaths	Countries Reporting Most Deaths
Africa	11,434,995	250,893	South Africa (99,725), Tunisia (28,062), Egypt (24,277), Morocco (16,039), and Ethiopia (7,486)
Asia	109,376,701	1,246,923	India (515,877), Indonesia (152,437), Iran (138,949), Philippines (57,625), and Vietnam (41,477)
America	149,518,189	2,667,857	United States (965,105), Brazil (655 249), Mexico (321,115), Peru (211,579), and Colombia (139,315)
Europe	183,944,230	1,883,711	Russia (361,344), United Kingdom (163,079), Italy (156,868), France (153,824), and Germany (125,856)
Oceania	3,904,300	8,632	Australia (5,591), Fiji (834), French Polynesia (645), Papua New Guinea (639), and Guam (338)
Other	705	6	Deaths reported from an international conveyance in Japan

Sources: European Centre for Disease Prevention and Control (March 2022) and Attard and Mulley (2022).

Later data are less reliable as 'living with COVID-19' meant that fewer people were testing, and so the number of cases is probably much under-recorded.

TRANSPORT BEHAVIOUR IMPACTS OF THE PANDEMIC DISRUPTION

Transport generally was indeed a vector of contagion. The spread over nations and the speed at which this occurred can be linked to both transport policy and the proximity to transport hubs. In the United States, Zare et al. (2022) demonstrated how the speed and spatial spread of the COVID-19 virus was positively associated with the number of airports and their level of activity (see also Jenelius, 2022; Kent & Crane, 2022; Lättman et al., 2023; Nguyen et al., 2021; Oviedo et al., 2022; Palm, 2022). Although passenger air transport came to a halt at the start of the pandemic (see Budd et al., 2020), the post-pandemic rebound in passenger numbers for air travel has outstripped many forecasts. According to the first United Nations World Tourism Observatory (UNWTO, 2024) *World Tourism Barometer*, tourism was set to reach pre-pandemic levels in 2024 despite ongoing challenges for the air sector that include staff shortages, high oil prices, and inflation.

Several studies focused on public transport impacts, including the measures undertaken by authorities and operators to secure the viable operation of services during and after pandemic conditions (Tirachini & Cats, 2020). Public transport is particularly important for just and sustainable mobility, but the risks associated with contagion undermined operations to the extent that emergency funding was needed in the face of plummeting patronage (although in some cases the decline in patronage was from an already declining base (Palm, 2022)). Especially impactful in some places was the cessation of many commuting trips because only essential workers were allowed to travel, and the degree of decline was related to the severity of the lockdown (Van Acker, 2022). Research also uncovered how different demographic groups suffered different levels of exposure to COVID-19 as a result of their transport needs: those dependent on public transport had to travel in the lockdowns because of work or were the first to return for non-work travel when restrictions started to ease (Amin & Uredo Adah, 2022; Marra et al., 2022).

Public transport services quickly adopted new measures to improve health control in vehicles and reduce virus anxiety (Jenelius, 2022), including temperature screening, tracking apps, mask wearing, social distancing, and frequent disinfection and improvements in vehicle air quality through the installation of filters in air conditioning units. Other complementary measures also indirectly affected public transport, among them working-from-home (WFH) policies which removed the need for commuting to work, removal of parking to increase space for walking and distancing, and in some cases the implementation of pop-up cycle lanes which took away even more space from public transport (see, for example, Nelson et al., 2022). Chen et al. (2024) studied the effects of proactive and reactive health control measures implemented for public transport in Hong Kong, a city with a high percentage of population dependent on public transport.

Their stated preference survey found that health controls were appreciated by passengers but that the effect was moderate in terms of encouraging a return to public transport use post-lockdown.

Alongside these disruptive changes were positives arising from changes in travel behaviour. Lockdowns affected private car travel, which fell drastically during the first few months of the pandemic because of forced interventions by public health authorities. The most significant benefits were in terms of congestion and emissions, with many cities across the world experiencing a dramatic decrease in pollution levels. Indeed, the COVID-19 pandemic showed what cities would look like without traffic and pollution (see Malpede & Percoco, 2020; Nieuwenhuijsen, 2021; Slezakova & Pereira, 2021). In some, the reduction in pollution led to the clearer visibility of surrounding environments, such as in Punjab, India, where the peaks of the Himalayas 'could be seen for the first time in decades' (Picheta, 2020). Perhaps more importantly, the reduction in car travel, combined with lockdown and physical distancing rules, stimulated an array of infrastructural provisions for active travel, from the Open Streets and Slow Streets programmes in New York and San Francisco (see Chapter 3) to the extension of cycling networks in Bogota and Mexico City and 'pop up' cycle routes in many cities such as Sydney and London. Combs (2024) recorded over 900 interventions in support of walking and cycling in over 400 cities in an online database (see also Buehler & Pucher, 2024; Nelson et al., 2022; Nikitas et al., 2021).

The need to ensure social distancing also led to newer modes of travel. Sharing services offered by transport network companies (TNCs), often referred to as ride-hailing companies, grew in popularity, with consequential health and safety changes being made to protect users, passengers, and drivers (Daus, 2022). These were accompanied by a rise in individual micromobility with the further spread of shared e-scooter, e-bike, and bicycle schemes (Attard & Balbontin, 2024; Goh et al., 2023). In this case, the technology providers capitalised on both the circumstances of the pandemic and the pervasive and extensive use of mobile smartphones to offer services capable of revolutionising urban travel. In July 2023, e-scooter operator Bolt launched the sixth-generation e-scooter, with more sensors, microprocessors, data mining capabilities, and precision technologies that would achieve new benchmarks for safety, sustainability, and user experience (Bolt Blog, 2023). With WFH becoming common practice, in the first few months of lockdown, relatively 'old' technologies allowing online video conferencing sprung to the forefront and were upgraded to handle surging demand. Further tools and services were developed to facilitate more and more activities online, including online conference platforms, e-learning, and e-shopping.

While some of the observed changes in travel behaviour in cities were temporary, some appear to have withstood the test of time and become more permanent in nature (see Molloy et al., 2021; Shakibaei et al., 2021). Working patterns, particularly for white-collar workers, appear to have withstood the test of time, and WFH has become established for at least part of the working week. This has had consequential changes to business requirements for office space and potentially significant changes on home location choices and home space requirements (Cui & Attard, 2024; Mariotti et al., 2022). In terms of emission reductions, WFH is a positive change since it reduces the absolute amount of commuting with its associated emission patterns during peak hours. However, there are other implications which may be less favourable, including a decline in office culture and, for cities with central business districts, a decline in the vitality of those areas. Whether these changes in spatial patterns lead to positive or negative outcomes in the longer terms remain to be seen.

Freight and supply chains were also affected. Lockdowns triggered not only a change in consumption patterns but also a rather sudden rise in online shopping, and this in turn affected the distribution of goods. The disruption led to delivery delays, again unevenly spread depending on a particular country's manufacturing base and location. The effect of the pandemic on the freight sector was mixed, with growth bringing the need to adapt to new ways of working (Beckers & Cant, 2024; Kiani Mavi et al., 2022). The rise of small urban deliveries encouraged the growth of new economic operators delivering packages in different ways. There is no doubt that this increased activity heightens the impact on cities' heavily burdened (particularly road) infrastructures (see Cavallaro & Nocera, 2022), although an encouraging development has been the rise in cargo bike (or e-cargo bike) deliveries which offer a greener alternative to car- and van-based options (see, for example, Gonzalez-Calderon et al., 2022). Drones (Yuen et al., 2024), autonomous robots (Pani et al., 2020), and certain food delivery platforms (Wang et al., 2022) are also examples of innovations that offer alternatives to cars and vans post-pandemic.

DISRUPTION AND ITS IMPLICATIONS FOR DECARBONISATION

The previous sections have provided an overview of the many studies that have investigated change in transport actions, policy, and behaviour during and after the COVID-19 pandemic. Is there the potential for some of these to bring benefits to global climate change efforts? There have been changes in travel behaviour and transport policies which have persisted, and these have had both positive and negative impacts on climate change mitigation. This section discusses the research which has investigated the impacts, perceptions, acceptance, and forecast implications for the future and in so doing raises questions relating to resilience as well as highlighting the importance of good governance in achieving change.

Marsden and Docherty (2021) were among the first to look at the issue of governance during COVID-19. They considered the processes of policy formulation and the institutions that implement measures following the rapid response of the UK government in the transport sector, with findings pointing towards an accelerated response to already planned policies but with few new ideas of how to tackle future demand. Unfortunately, the concept of the new, post-pandemic normal (Attard & Mulley, 2022) which embraces the need for radical policy change to mitigate and adapt to climate change seems to have been sidelined: we have seen (rapidly) rising traffic levels in cities across the world (INRIX, 2021, cited in Attard, 2023), increasing tourism travel particularly by air and sea, and a reluctance in some cities to pursue measures that are more sustainable (Scerri et al., 2024). Kyriakidis et al.'s (2023) discussion of Athens' experience is an example of why this reluctance can come about. The city's attempt to implement the Great Walk of Athens – a high-profile, sustainable transport intervention package – produced mixed results, leading to increased traffic delays and only marginal walking and cycling uptake. Citizen resistance arose due to a lack of public consultation in the decision-making process. It is a case study about reactive, potentially rushed, redevelopment measures which are difficult to build and retain.

The impacts of climate change on the various aspects of the transport systems across different geographies have been studied by many authors (see, for example, Chinowsky et al., 2015; Dawson et al., 2016; Espinet et al., 2016; Jaroszweski et al., 2010; Koetse & Rietveld, 2009; Qiao et al., 2022; Rattanachot et al., 2015; Ryley & Chapman, 2012; Schweikert et al., 2014; Wang, Qu, Yang, & Ng, 2020). The pressures on infrastructures brought about by climate events highlight the importance of addressing issues of vulnerability in transport systems and the policy response required to upgrade infrastructure to withstand climate pressures but also address resilience in responding to climate incidents. The opportunity of investigating the effects of disruptive events (like COVID-19) and the subsequent actions to quickly roll out effective measures could help with decarbonisation efforts.

Another aspect is that of technologies that flourished during the pandemic to support less-mobile lifestyles. While these technologies have allowed for work and study to be conducted online, and shopping to be ordered and delivered remotely, research into the different pathways of practices, policy, and implementation is required to investigate how they are likely to change transport and mobility across different geographies and communities in the longer run. There is potentially much to gain, especially if it becomes easier to embrace concepts such as the 15-minute city (Moreno et al., 2021) and implement proximity planning (Silva et al., 2023) which goes a long way to managing travel demand with the potential to improve sustainability.

The movement of goods across the world and in our cities remains a challenging issue both in meeting climate change goals and promoting sustainability. The COVID-19 pandemic did little to curb the growth in consumption and movement of goods globally. And recent events have reminded us of the dependence of global trade on a transport system which is prone to disruptions. When one of the largest container ships in the world, the Ever Given, ran aground in the Suez Canal in March 2021 and blocked the passage of vessels for six days, havoc ensued in world maritime trade. By the sixth day, almost 400 container ships were queuing to use the canal. Furthermore, in February 2022, Russia invaded Ukraine, leading to the largest conflict in Europe since World War II; this was followed by the war in the Gaza Strip and Israel. Both conflicts are still ongoing at the time of writing. Such disruptions continue to have predominantly negative impacts on the movement of people and goods, but it is also difficult to operationalise effective climate change mitigation policies in places of conflict, and as indicated above, climate change is a global phenomenon.

CONCLUSION

In any consideration of the potential for disruptive events to aid in making progress towards net zero efforts, an important principle is that of equity and justice. Care needs to be taken so that inequalities within and between nations are not exacerbated by measures that are taken to decarbonise; here the outcome from some of the COVID-19 measures, where the impacts were differentially felt by different socio-economic groups in different places, should not be repeated. Accessibility was restricted for lower economic socio-demographic groups, and as the world moved to 'living with COVID', it is these same groups who continue to struggle to work from home or carry out online shopping because of the nature of their work, residential location, or socio-demographic characteristics. As such, they struggle to reduce their dependence on essential travel services. At the same time, lower public transport patronage, primarily fuelled by a lower commuting need, might encourage operators to reduce their peak vehicle requirements and reduce service availability for those most dependent upon it. Strategies to promote mode shift by wooing discretionary travellers on to public transport thus remain paramount.

The ideal of maintaining and delivering growth while also reducing consumption and limiting emissions is possible through the lessons learnt from the COVID-19 pandemic recovery. Due in part to other events – wars, political instability, and trade disruptions – that have attracted global attention away from the climate challenge, there is an urgency to look back and understand the potential of growth through more sustainable mobility. In addition to this, the COVID-19 pandemic has demonstrated the power of unity, across borders and continents. If the fight for decarbonisation becomes action on a truly global scale, we should see that reflected in increased success. Challenges such as those experienced during the pandemic can be overcome through harmonisation of principles and values, support and integration between economies, and building structures which allow for understanding and growth in overcoming the challenges of climate change. This includes the way in which countries need to be persuaded on the need to implement tough and difficult decisions capable of addressing net zero challenges.

COVID-19 benefited from unprecedented global co-operation necessary because of the speed of contagion and the impact of transport as a vector of transmission. Climate change is happening more slowly – or, at least, it has been present for some time but is becoming increasingly more urgent. The governance processes necessary for the delivery of climate action need to be formulated in a similar spirit of global cooperation as those experienced during the COVID-19 pandemic, as the world comes under increasing pressure to take steps to curtail emissions and decarbonise more generally. The United Nations Conference of the Parties (COP) has a role in identifying and supporting winners and losers in the adoption of mitigation and adaptation measures and ensuring the understanding that if we manage to slow down (and maybe reverse) carbon emissions, everyone will stand to benefit in the long term. In all of this, transport has an important role to play because of its contribution to the overall emissions burden. Knowledge sharing and transfer of practices are required to make the process faster and more effective.

Ultimately, the issues we are facing due to climate change are not new but are ones that could lead to outcomes more severe than the COVID-19 pandemic. Unlike the pandemic, however, there is no vaccine that would see a slow but sure recovery for the planet and its inhabitants.

REFERENCES

Amin, S., & Uredo Adah, J. (2022). COVID-19 influence on commuters' attitude towards riding public buses for essential trips. *Cities*, *131*, 103890.

Association for Professionals in Infection Control and Epidemiology. (2014, April 23). *Infection protection and you*. Retrieved May 10, from https://apic.org/Resource_/TinyMceFileManager/for_consumers/IPandYou_Bulletin_Outbreaks_Epidemics_Pandemics.pdf

Attard, M. (2023). Planning for bus rapid transit in an island context. The challenges of implementing BRT in Malta. *Research in Transportation Economics*, *100*, 101309.

Attard, M., & Balbontin, C. (2024). Workshop 6 report: Micromobility movement in urban transport. *Research in Transportation Economics*, *103*, 101399.

Attard, M., & Mulley, C. (2022). Transport and pandemic experiences: An introduction. In M. Attard & C. Mulley (Eds.), *Transport and pandemic experiences* (pp. 3–13). Emerald Publishing Limited

Beckers, J., & Cant, J. (2024). Half a decade in two years: Household freight after COVID-19. *Transport Reviews*, *44*(2), 544–565.

Bešinović, N. (2020). Resilience in railway transport systems: A literature review and research agenda. *Transport Reviews*, *40*(4), 457–478.

Bolt Blog. (2023). *Safer. Smarter. Stronger*. Meet the new Bolt 6 scooter. https://bolt.eu/en/blog/introducing-bolt-6-scooter/#:~:text=We're%20excited%20to%20introduce,%2C%20sustainability%2C%20and%20user%20experience

Budd, L., Griggs, S., Howarth, D., & Ison, S. G. (2011). A fiasco of volcanic proportions? Eyjafjallajokull and the closure of European Airspace. *Mobilities*, *6*(1), 31–40.

Budd, L., Ison. S. G., & Adrienne, N. (2020). European airline response to the COVID-19 pandemic – Contraction, consolidation and future considerations for airline business and management. *Research in Transportation Business and Management*, *37*, 100578.

Buehler, R., & Pucher, J. (2024). COVID-19 and cycling: A review of the literature on changes in cycling levels and government policies from 2019 to 2022. *Transport Reviews*, *44*(2), 299–344.

Cavallaro, F., & Nocera, S. (2022). Integration of passenger and freight transport: A concept-centric literature review. *Research in Transportation Business and Management*, *43*, 100718.

Chen, T., Fu, X., Hensher, D. A., Li, Z.-C., & Sze, N. N. (2024). Effects of proactive and reactive health control measures on public transport preferences of passengers – A stated preference study during the COVID-19 pandemic. *Transport Policy*, *146*, 175–192.

Chinowsky, P. S., Schweikert, A. E., Strzepek, N., & Strzepek, K. (2015). Road infrastructure and climate change in Vietnam. *Sustainability*, *7*(5), 5452–5470.

Combs, T. (2024). *Local actions to support walking and cycling during social distancing dataset*. http://pedbikeinfo.org/resources/resources_details.cfm?id=5209

Creutzig, F., Roy, J., Lamb, W. F., Azevedo, I. M. L., Bruine de Buin, W., Dalkmann, H., Edelenbosch, O. Y., Geels, F. W., Brubler, A., Hepburn, C., Hertwich, E. G., Khosla, R., Mattauch, L., Minx, J. C., Ramakrishnan, A., Rao, N. D., Steinberger, J. K., Tavoni, M., Ürge-Vorsatz, D., & Weber, E. U. (2018). Towards demand-side solutions from mitigating climate change. *Nature Climate Change*, *8*(4), 268–271.

Cui, J., & Attard, M. (2024). COVID-19 and short-term housing: Economic and social impacts and implications. *Journal of Housing and the Built Environment*, *39*, 1103–1128. https://doi.org/10.1007/s10901-024-10117-6

Daus, M. W. (2022). Development of policies for ridesourcing and taxis. In M. Attard & C. Mulley (Eds.), *Transport and pandemic experiences* (pp. 203–220). Emerald Publishing Limited.

Dawson, D., Shaw, J., & Roland Gehrels, W. (2016). Sea-level rise impacts on transport infrastructure: The notorious case of the coastal railway line at Dawlish, England. *Journal of Transport Geography*, *51*, 97–109.

Espinet, X., Schweikert, A., van den Heever, N., & Chinowsky, P. (2016). Planning resilient roads for the future environment and climate change. Quantifying the vulnerability of the primary infrastructure system in Mexico. *Transport Policy*, *56*, 78–86.

Goh, H., Choi, G., & Song, Y. (2023). How the COVID-19 pandemic changed travel behaviour? A case study on public bikes in Seoul. *Research in Transportation Economics*, *101*, 101339.

Gonzalez-Calderon, C. A., Posada-Henao, J. J., Granada-Muñoz, C. A., Moreno-Palacio, D. P., & Arcila-Mena, G. (2022). Cargo bicycles as an alternative to make sustainable last-mile deliveries in Medellin, Colombia. *Case Studies on Transport Policy*, *10*(2), 1172–1187.

Intergovernmental Panel on Climate Change. (2021). *IPCC AR6 WGI: Summary for policymakers*. www.ipcc.ch/report/ar6/wg1/downloads/report/IPCC_AR6_WGI_SPM.pdf

Jaroszweski, D., Chapman, L., & Petts, J. (2010). Assessing the potential impact of climate change on transportation: The need for an interdisciplinary approach. *Journal of Transport Geography*, *18*(2), 331–335.

Jenelius, E. (2022). Impact on public transport. In M. Attard & C. Mulley (Eds.), *Transport and pandemic experiences* (pp. 287–302). Emerald Publishing Limited.

Kent, J. L., & Crane, M. (2022). Impacts on health. In M. Attard & C. Mulley (Eds.), *Transport and pandemic experiences* (pp. 303–322). Emerald Publishing Limited.

Kiani Mavi, R., Kiani Mavi, N., Olaru, D., Biermann, S., & Chi, S. (2022). Innovations in freight transport: A systematic literature evaluation and COVID implications. *The International Journal of Logistics Management*, *33*(4), 1157–1195.

Koetse, M. J., & Rietveld, P. (2009). The impact of climate change and weather on transport: An overview of empirical findings. *Transportation Research Part D: Transport and Environment*, *14*(3), 205–221.

Kyriakidis, C., Chatziioannou, I., Iliadis, F., Nikitas, A., & Bakogiannis, E. (2023). Evaluating the public acceptance of sustainable mobility interventions responding to COVID-19: The case of the Great Walk of Athens and the importance of citizen engagement. *Cities*, *132*, 103966.

Lättman, K., Olsson, L. E., Waygood, O. D., & Friman, M. (2023). Nowhere to go – Effects on elderly's travel during COVID-19. *Travel Behaviour and Society*, *32*, 100574.

Malpede, M., & Percoco, M. (2020). *Lockdown measures and air quality: Evidence from Italian provinces*. SSRN. https://ssrn.com/abstract=3714673

Mariotti, I., Di Marino, M., & Bednář, P. (2022). Introduction. In I. Mariotti, M. Di Marino, & P. Bednář (Eds.), *The effects of the COVID-19 pandemic on the future of working spaces* (pp. 1–5). Routledge.

Marra, A. D., Sun, L., & Corman, F. (2022). The impact of COVID-19 pandemic on public transport usage and route choice: Evidences from a long-term tracking study in urban area. *Transport Policy*, *116*, 256–268.

Marsden, G., Anable, J., Chatterton, T., Docherty, I., Faulconbridge, J., Murray, L., Roby, H., & Shires, J. (2020). Studying disruptive events: Innovations in behaviour, opportunities for lower carbon transport policy? *Transport Policy*, *94*, 89–101.

Marsden, G., & Docherty, I. (2021). Mega-disruptions and policy change: Lessons from the mobility sector in response to the Covid-19 pandemic in the UK. *Transport Policy*, *110*, 86–97.

Molloy, J., Schatzmann, T., Schoeman, B., Tchervenkov, C., Hintermann, B., & Axhausen, K. W. (2021). Observed impacts of the COVID-19 first wave on travel behaviour in Switzerland based on large GPS panel. *Transport Policy*, *104*, 43–51.

Monios, J., & Wilmsmeier, G. (2020). Deep adaptation to climate change in the maritime transport sector – A new paradigm for maritime economics? *Maritime Policy & Management*, *47*(7), 853–872.

Moreno, C., Allam, Z., Chabaud, D., Gall, C., & Pratlong, F. (2021). Introducing the "15-minute city": Sustainability, resilience and place identity in future post-pandemic cities. *Smart Cities*, *4*, 93–111.

Nelson, J. D., Clifton, G., & Loyola, M. (2022). Policies for public transport. In M. Attard & C. Mulley (Eds.), *Transport and pandemic experiences* (Vol. 17, pp. 185–202). Emerald Publishing Limited.

Nguyen, M. H., Pojani, D., Nguyen T. C., & Ha, T. T. (2021). The impact of Covid-19 on children's active travel to school in Vietnam. *Journal of Transport Geography*, *96*, 103191.

Nieuwenhuijsen, M. J. (2021). New urban models for more sustainable, liveable and healthier cities post covid19; reducing air pollution, noise and heat island effects and increasing green space and physical activity. *Environment International*, *157*, 106850.

Nikitas, A., Tsigdinos, S., Karolemeas, C., Kourmpa, E., & Bakogiannis, E. (2021). Cycling in the era of COVID-19: Lessons learnt and best practice policy recommendations for a more bike-centric future. *Sustainability*, *13*, 4620.

Oviedo, D., Guzman, L. A., Arellana, J., Sabogal-Cardona, O., Moncada, C., & Scholl, L. (2022). Cities under lockdown: Mobility and access inequalities stemming from COVID-19 in urban Colombia. In M. Attard & C. Mulley (Eds.), *Transport and pandemic experiences* (pp. 107–126). Emerald Publishing Limited.

Palm, M. (2022). North American transportation during COVID-19: What really changed? In M. Attard & C. Mulley (Eds.), *Transport and pandemic experiences* (pp. 55–75). Emerald Publishing Limited.

Pani, A., Mishra, S., Golias, M., & Figliozzi, M. (2020). Evaluating public acceptance of autonomous delivery robots during COVID-19 pandemic. *Transportation Research Part D: Transport and Environment*, *89*, 102600.

Picheta, R. (2020, April 9). *People in India can see the Himalayas for the first time in 'decades', as the lockdown eases air pollution*. CNN. https://edition.cnn.com/travel/article/himalayas-visible-lockdown-india-scli-intl/index.html

Qiao, Y., Guo, Y., Stoner, A. M. K., & Santos, J. (2022). Impacts of future climate change on flexible road pavement economics: A life cycle costs analysis of 24 case studies across the United States. *Sustainable Cities & Society*, *80*, 103773.

Rattanachot, W., Wang, Y., Chong, D., & Suwansawas, S. (2015). Adaptation strategies of transport infrastructures to global climate change. *Transport Policy*, *41*, 159–166.

Rejan, K., & Ahn, Y. (2024). Challenges, experiences, and coping behaviors among SMEs in the adventure tourism industry after the COVID-19 pandemic: A case from Pokhara, Nepal. *Journal of Outdoor Recreation and Tourism*, *47*, 100797.

Ritchie, H. (2020). *Cars, planes, trains: Where do CO_2 emissions from transport come from?* https://ourworldindata.org/co2-emissions-from-transport

Ryley, T., & Chapman, L. (2012). *Transport and climate change*. Emerald Publishing.

Scerri, K., Martinez, L., Betancur Arenas, J., Keserü, I., & Attard, M. (2024, July 3–5). *Understanding the barriers and facilitators of designing and implementing active mobility intervention in Brussels and Malta*. Presentation at the 17th international NECTAR conference, Brussels.

Schweikert, A., Chinowsky, P., Kwiatkowski, K., & Espinet, X. (2014). The infrastructure planning support system: Analyzing the impact of climate change on road infrastructure and development. *Transport Policy*, *35*, 146–153.

Shakibaei, S., de Jong, G. C., Alpkökin, P., & Rashid, T. H. (2021). Impact of the COVID-19 pandemic on travel behaviour in Istanbul: A panel data analysis. *Sustainable Cities and Society*, *65*, 102619.

Silva, C., Büttner, B., Seisenberger, S., & Rauli, A. (2023). Proximity-centred accessibility – A conceptual debate involving experts and planning practitioners. *Journal of Urban Mobility*, *4*, 100060.

Slezakova, K., & Pereira, M. C. (2021). 2020 COVID-19 lockdown and the impacts on air quality with emphasis on urban, suburban and rural zones. *Scientific Reports*, *11*, 21336.

Stern, N. (2007). *The economics of climate change: The Stern review*. Cambridge University Press.

Stojčić, N., & Vizek, M. (2024). Did COVID-19 grants support tourism firms in securing survival and employment? Evidence from Croatia. *Tourism Management*, *104*, 104934.

Tirachini, A., & Cats, O. (2020). COVID-19 and public transportation: Current assessment, prospects, and research needs. *Journal of Public Transportation*, *22*(1), 1–21.

United Nations World Tourism Observatory. (2024). *International tourism to reach pre-pandemic levels in 2024*. https://www.unwto.org/news/international-tourism-to-reach-pre-pandemic-levels-in-2024

Van Acker, V. (2022). Changes in travel behaviour in Europe. In M. Attard & C. Mulley (Eds.), *Transport and pandemic experiences* (pp. 77–106). Emerald Publishing Limited.

Van Exel, N., & Rietveld, P. (2009). When strike comes to town … anticipated and actual behavioural reactions to a one-day, pre-announced, complete rail strike in The Netherlands. *Transportation Research Part A: Policy and Planning*, *43*, 526–535.

Wang, T., Qu, Z., Yang, Z., Nichol, T., Clarke, G., & Ge, Y. E. (2020). Climate change research on transportation systems: Climate risks, adaptation and planning. *Transportation Research Part D: Transport and Environment*, *88*, 102553.

Wang, T., Qu, Z., Yang, Z., & Ng, A. K. Y. (2020). Chapter 6 – How does the UK transport system respond to the risks posed by climate change? An analysis from the perspective of adaptation planning. In A. K. Y. Ng, J. Monios, & C. Jiang (Eds.), *Maritime transport and regional sustainability* (pp. 85–106). Elsevier.

Wang, X., Zhao, F., Tian, X., Min, S., von Cramon-Taubadel, S., Huang, J., & Fan, S. (2022). How online food delivery platforms contributed to the resilience of the urban food system in China during the COVID-19 pandemic. *Global Food Security*, *35*, 100658.

Yuen, K. F., Cai, L., Wee, S. C. K., & Wang, X. (2024). Pandemic-driven acceptance of urban drone deliveries. *Transport Policy*, *146*, 356–370.

Zare, H., Delarmente, B., & Gaskin, D. J. (2022). How do transportation policies drive geographic disparities in COVID-19 infections and deaths in the United States? In M. Attard & C. Mulley (Eds.), *Transport and pandemic experiences* (pp. 39–51). Emerald Publishing Limited.

INDEX

Printed and bound by CPI Group (UK) Ltd, Croydon, CR0 4YY

15/10/2025

14753499-0002